"To Great Aunt Helen

Courtney Hope

WITCH WAY
PUBLISHING

First Edition, 2023

Witch Way Publishing
3436 Magazine Street
#460
New Orleans, LA 70115
www.witchwaypublishing.com

Copyright © 2023 by Courtney Hope

Editor: Tonya Brown
Copy Editor: Anna Rowyn
Cover Designer: Quirky Circe Designs
Illustrator: Haley Newman

Printed in the United States of America

ISBN Paperback: 9781088186664
ISBN E-Book: 9781088186732

The Hygge Witch Handbook

Courtney Hope

Courtney Hope

Table of Contents

Courtney Hope

hygge

[ˈhyːgə] noun

the practice of creating cosy and congenial environments
that promote emotional well-being.

FORWARD

My name is Courtney Hope and I am an author, a practicing witch, a modern day Persephone, and a Halloween Queen. My background is in interior design and event management, and I have utilised the skills, knowledge and experience I've gained from those fields to create my own unique magical practice and style: the *Hygge Witch*.

I have always been a very curious, spiritual person in pursuit of *hygge*; of comforts and welcoming energies into my home, which I have discovered go hand-in-hand with my practices as a witch. I believe the environment that you surround yourself with helps create who you are as a person and the kind of magic you manifest into your life. That's why I created this handbook as a way to blend important spiritual health and wellness with the comforts of heart and hearth.

It is my aim within *The Hygge Witch Handbook* to detail the spells and magical practices aligned with cleansing, decorating, and protecting your home in the most *hygge* way possible, sharing some of my personal favourite conjuring and manifesting spells, kitchen witchery, interior design and decorating tips along the way. I've also included other important information I feel every witch should know - *hygge* or not!

I, like many witches before me, have determined my own form of spell work by exploring the knowledge of other modern witches, taking pearls of wisdom and combining them with the natural comforts of a *hygge* lifestyle. *The Hygge Witch Handbook* is suffused with my personalised home-grown magic; I hope you enjoy it, and that it helps you harness your magic to live well in this modern age.

INTRODUCTION

What is Hygge?

Even if you haven't heard of *hygge,* you've seen the trends it inspires. It seems every picture online depicts it as natural tones, layered chunky knits, and faux fur rugs. Comfy beds with artfully tossed throw blankets and cushions abound alongside timber tables crammed cosily full of plants and candles, stacks of books next to a roaring fire, and photos of people in cable knit jumpers with mugs full of steaming hot chocolate. The *hygge* aesthetic is the perfect antidote for those tired of the cold, sterile, and modern. But the concept of *hygge* is so much more than cute pictures of woollen socks and piles of books.

Hygge (pronounced hue-guh) is a Danish word for feeling cosy and comfortable, creating a sense of wellbeing and contentment in your environment. It is a cultural lifestyle that values the feeling of satisfaction evoked by the simple comforts of life: the warmth of a fire after a hard day's activities, the glow of good conversations with loved ones, the enjoyment of a fine full-bodied wine or silky cocoa, and finding a sense of peace in your environment.

Denmark – the birthplace of hygge – is considered one of the happiest countries on Earth by the European Social Survey, which records high levels of well-being and quality of life in the drizzly country. Although many people believe time spent indoors due to the inclement weather is why Nordic people created the hygge lifestyle, hygge is really the creation of one's own micro-universe with space to recharge, entertain friends, enrich their lives, and enjoy emotional comforts – a valuable exercise whatever the weather.

In Denmark, hygge is ingrained in the culture, but it is exported to other parts of the world as the warmth and happiness that comes from enjoying your surroundings and the company you are keeping. Hygge exists in moments of contentment, expressed as a feeling of being warm, safe, comforted and sheltered.

This is the true spirit of hygge.

What is Witchcraft?

Witchcraft has been practised for centuries. Although the term originated in medieval Europe, it has been applied to beliefs and occult practises in many different cultures and countries. Indigenous communities on virtually every continent practise religions that contain elements of witchcraft and magical thinking; although witchcraft is open to everyone, some communities have closed practices, where only those invited into that particular religion and culture can practise that form of magic. When creating your magic, be respectful of cultural sensitivities.

In contemporary Western culture, modern pagans and witches can use witchcraft as part of a healing journey and wellness rituals, but each individual's practise can look completely different to another witch's, even within covens of like-minded practitioners or as part of a witchcraft religion.

Witchcraft really is something that you can make your own. While the word *witch* has a deep etymology and has been used pejoratively in the past, the label itself has changed with the times and is now seen in Western culture as a modernized identity.

With the rise of monotheism (belief in one singular deity), this likening of witchcraft to heresy and the worship of 'false gods' brought years of torment, suspicion, and lack of understanding onto people of the 14th century. Traditional practices and beliefs were maligned in works like the 1486 tome *The Malleus Maleficarum.* Many people were accused of witchcraft for 'suspicious' and 'unnatural' behaviour, mostly devout monotheists that did not practice any form of witchcraft. The numbers of the accused and the executed in Europe is staggering, with an estimation of up to 80,000 suspected witches that were put to death in Europe between 1500 and 1660, although some estimate that these numbers are much higher.

Magic in medieval Denmark and other Nordic countries, like most of continental Europe, was functional. Charms were used for protecting one from illness, predicting future weather, midwifery, and divination regarding the results of the upcoming harvest.

Some Danish witches found their way to the practice through natural religions and shamanism, with other beliefs stemming from Nordic mythology. Nordic mythology holds a very important role in witchcraft through Denmark.

During the Viking Age, the Norse people of Scandinavia began exploring and settling other territories around North-western Europe, taking Norse paganism with them. Although the Old Norse religion was gradually replaced by Christianity and had virtually disappeared by the 12th century, elements of Norse mythology continued into Scandinavian folklore.

Early rural Nordic laws, like most criminal laws about witchcraft, cited penalties against those that use magic to physically harm men, women, children, or animals. Known as *forgorning*, one such law entitled victims to retribution by killing the sorcerer, but was eventually replaced by a more formal death penalty before being revoked. Laws against witchcraft still exist in many countries around the world, and suspicions of witchcraft can still carry the death penalty today.

There are many aspects that can influence how a person practices or uses magic, including cultural and ethnic identity. There is no right or wrong way to be a witch or study witchcraft, and you don't have to be particularly religious to benefit from it. You can use your practice to connect with specific deities for assistance, practice the knowledge that has been sent down to you through history's pages, or you focus on positive outcomes to support your lifestyle.

As I learnt more about the rituals and history of witchcraft, I have realised that my practice is part of a modern lifestyle movement that uses rituals to set good intentions and raise my vibrations to a higher level. Through exploring different types of practices, I also realised that my passion for the concept of hygge was beginning to have a magical effect on my personality, self-worth, manifestations, and way of life.

This is how I found myself drawn into the idea of becoming a Hygge Witch.

What Is a Hygge Witch?

There are many names for a witch who practices her magic around her home, her garden, and her kitchen. A *Green Witch* is known to focus her practice in nature, tending and growing herbs and plants in her garden before utilising them in her spell work. She often has an impressive green thumb, are in cyclical symbiosis with the seasons, and love to forage in deep forest clearings. A *Kitchen Witch* creates powerful magic through cooking and baking, creating delicious feasts heaped with love and intention. A *Hedge Witch* practices herbalism, energy healing and counselling as well as natural mysticism – walking the 'hedge' between the natural and the spiritual realms.

Although the *Hygge Witch* practices elements of all these paths, the closest archetype to the Hygge Witch is the *Hearth Witch.* Hearth Witches focus their magical practice on their home and daily life, setting intentions and protections around all who dwell within. Although many aspects of the Hygge Witch are taken from the Hearth Witch, there is a key difference: Hearth Witch focuses more on the spells and activities *in* the home, rather than the spells and activities that reflect the energy *of* the home and those residing inside it.

If hygge is about the overall feeling and warmth of an environment and being present in the moment. and witchcraft is a nature-based practice all about harnessing the world and your environment to provide the energy and intention you seek, it makes perfect sense for the two principles to live within each other.

Practising hygge provides a focus on comfort, connection, positive energy, and wellbeing. This can enhance your witchcraft by developing awareness of your surroundings, connecting you to your community, and encouraging you to live your daily life infused with magic. Hygge weaves an old-world spiritual magic into the presence of everyday life, reminding you, in the words of hearth witch Ann Murphy-Hiscock, that 'the very living of your life is a spiritual act'. It takes the magic, power and energy from your surroundings and infuses it into your domestic tasks, lifestyle choices, and spell work.

Hygge witchcraft is the practice of turning the mundane into the magical; creating a positive energetic space in your home and your heart that comforts, nourishes, and observes the simple pleasures of everyday life.

Hygge Witchcraft does this through six main pillars:

- **Energy** – Honouring the balance of the home as a sacred space.
- **Security** – Protecting yourself and your loved ones from negative energies, ensuring safety and provision.
- **Connection** – Connecting to oneself, to others, to the environment and to your shelter.

- **Wellness** – Supporting happiness and health for those that dwell in your home.
- **Comfort** – Creating a magical hearth that nurtures and comforts your soul.
- **Ritual** – Observing every day.

The next pages of *The Hygge Witch Handbook* will take you through these six different pillars of hygge, touching on different lifestyle designs, spells, and practices, as well as techniques that will raise your vibration of your home and create a perfect hygge environment - more than just a hard-to-pronounce word or a cute photo on Instagram.

THE 6 PILLARS OF BEING A

Hygge Witch

Energy

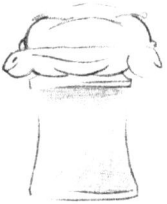

Security

Connection

Wellness

Comfort

Ritual

Courtney Hope

ENERGY

Understanding the energy of your surroundings is an important first step in practising Hygge Witchcraft. It will help you acknowledge, strengthen, and honour your home.

There are many different 'types' of energy, including physical, mechanical, spiritual, mental, and kinetic. Scientifically, all objects are made of atoms (which are made of energy); the energy stored within an object due to the position, arrangement or state of that object is known as *potential energy*.

The potential energy of objects in witchcraft isn't mechanical or physical, but metaphysical. Metaphysical is a term used to define the organic life force of something beyond the physical, unable to be touched or seen with the human eye. There are many different esoteric and metaphysical philosophies that work with this kind of energy.

One of these philosophies is part of the Wiccan religion called the 'Wiccan Rede', a statement of practise that provides the key moral systems of wiccan-based faiths and paths. One of the most key parts of the Rede, "an ye harm none, do what you will", highlights the idea of the three-fold law. The three-fold law is the idea that whatever energy you put into the world, whether positive or negative, will be returned to you three-fold.

The three-fold law is similar to the Buddhist law of *Karma*, in which you get back whatever energetic *intentions* you put out into the world. An *intention*, 'an aim or a plan', is connected to a manifestation; what you intend is indeed what you manifest. In other words, what you think about or focus on eventually occurs.

This focus on positive intention and manifestation is something that some witches naturally practice. Both the 'Wiccan Rede' and the philosophy of *Karma* are good ways to demonstrate the idea that everything has a balance and a vibration, including items within your home, different materials you work with, and the intentions and acts of service you put into the world. What you put out you get back, including peace and security within the walls of your home.

Everything has a vibration. Hygge witchcraft utilises the home's energy to reach the levels of contentment we are seeking as Hygge Witches.

Consider *Feng Shui,* the Chinese philosophy-based practice that looks at how we can live in harmony with the principles of the natural world within our built environments. This is an example of how interior design affects the way we manifest things into our lives.

The words Feng Shui literally translate to the words 'wind' and 'water', reflecting the idea that human life can be optimised by connecting to and living in flow with *tao*, the environment around us. Tao is the way of nature, and the principles of Feng Shui follow the natural world, creating an environment using tools like a commanding position, the *bagua,* and the five elements of earth, air, fire, water, and metal.

The art of Feng Shui begins with defining a space's commanding position, the position in the room that most empowers you to use the space. The *bagua,* an energy map that represents parts of your life including wealth, reputation, partnership, family, children, knowledge, and career, is then overlayed onto a diagram of the space. When creating a space with Feng Shui, you need to identify parts of your life that require work, using the representations of that area – usually a shape, colour, season, number, and element - in the correct section of the room to activate the energy.

Activating the energies in particular areas of your rooms is common in modern interior decorating practices, with many homes reinvigorated by a burst of Feng Shui during its introduction to Western design and recent popular resurgence (although it is an ancient art that has been in practice for over 3,000 years), and through the introduction of popular 'tidying expert' Marie Kondo's *KonMari* philosophy of household order.

Marie Kondo's methods rely on the idea that your space has energy and you need to use it accordingly. The most well-known aspect of her method is keeping only items that 'spark joy'. Marie regularly taps and handles items to 'wake up' their energy while tidying them, conducting house blessings and using a tuning fork prior to her work. Her method is mindful, introspective, and forward-thinking and her philosophy aligns wonderfully with the spirit of hygge.

The best way to determine the energy in your home as a Hygge Witch is by engaging closely with your space through tidying and organising, cleansing the home of unwanted energies and objects that no longer provide comfort. There are many different ways that you can reset the energy of your home.

I have listed different methods of space clearing utilising the elements of earth, air, fire, and water (pg. 63). It's best, before beginning to cleanse the spirit and energy of the house, to cleanse the *actual* house first.

And yes, I am talking about a complete overhaul of your home and the things in it. I know it sounds like a momentous task, but cleaning thoroughly is an important way to explore the energies you are currently surrounding yourself with in your home.

There are many different unique tidying and organising approaches that really work wonders in a hygge home. In order to cleanse your home and reset the energies within it, before bringing new items and energies into your space, the best place to start is decluttering, tidying and organising the items already in your home.

By ensuring you clean out and organise your home properly, your clutter will never have the opportunity to change the positive vibration of your home into a negative one.

It's important to note that decorating, cleansing, and changing the vibrational energy of your home is not static. As we ebb and flow with our natural environment and the seasons of our lives, the energy within our home changes as well . Items are moved, replaced, and upcycled as we move through the energy of different seasons, moon cycles, and life moments. You as a person change constantly, and your home will change with you, but if you set yourself up to be as organised and tidy as possible it will make shifting the vibrations of your home to align with the seasons and moments of your life easier.

Mentioned briefly, one of my favourite methods for tidying is Marie Kondo's *KonMari* tidying philosophy. Usually tidying methods advocate room-by-room or little-by-little approaches, but the *KonMari Method* encourages tidying by category rather than location. Marie is specific with the order of the categories, beginning with clothes and then moving on to books, papers, the miscellaneous items that she refers to as 'komono', and finally ending with sentimental items. With her method of thanking the items for their service and letting them go, you can clear away the clutter to better live the life you want.

As someone who has always cleaned and tidied by areas, I have found going by item and not location is a much easier way to take stock of all the objects you have collected throughout your house: all the books you have on bedside tables and in reading nooks, the blankets you have stored in cupboards, and the leftover paperwork you haven't thrown away, even if it's in different locations.

Another one of my favourite ways to organise the items in my home is to live by the golden item to space ratio of 80:20. The idea of the ratio is that when displaying and storing items in your home, the space is only about 80% full, leaving 20% breathing room or neutral space around the items.

This strategy is commonly found in interior design but utilising the breathing room of the space is just as important when it comes to storage. Maintaining the 80:20 ratio stops things from looking too cluttered or disorganised and serves as a reminder for you to discard unused or unnecessary items if you run out of room.

Another great way to know when you have exceeded the designated space and need to go through your items is containing them in proper storage. It also beautifully and neatly displays your objects. As a Hygge Witch, my favourite way to organise my home is uniform storage containers and baskets, especially environmentally friendly and naturally spirited options like soft woven baskets, rigid wire, or natural fibrebaskets, and bamboo boxes. Baskets are an easier way to access the items that can otherwise get hidden at the back of the cupboard, and the hygge feeling of natural material storage solutions brings warmth and softness to my space.

Notes on Discarding

Many of my personal values and beliefs are based my experience that when you connect to nature, you come to see the balance in it. I believe in living an ethical life as a conscious consumer, vegan and environmentalist, and something that developed hand-in-hand with my journey into hygge and witchcraft. I noticed that a lot of tidying and organising methods don't touch on how to discard items properly once they no longer have a place in your home. Because of this, I have come up with some handy environmentally friendly ways to discard your unwanted items.

Your trash is another person's treasure!

When you go through your discard pile, it is best to further separate it into different options, always choosing the most environmentally friendly. Begin with labelling one pile the 'Sell' pile and one the 'Donate' pile for items of useable quality and condition - things you don't need any more but are too good to send to the tip. I get rid of these items by selling them on online marketplace platforms or 'Buy Nothing' groups, giving them away to that want them, and donating to a charity in my local area. A bit of research on what is near you can find excellent options for getting rid of specific items: schools will often take stationery and art supplies, animal shelters need old bedding and towels, and Little Street Libraries are a great way to leave your unwanted books for others to enjoy!

Reuse

It's often possible to repurpose something you already own before you move on to recycling it. Old clothing can be dyed, mended, or repurposed as raw materials for other projects, while unwanted food containers can be used in the garden to start seeds and grow plants. Many items can be incorporated into attractive, functional, and unique pieces for your home: a few hours of mindful crafting or a lick of paint can rejuvenate old furniture, make a beautiful rag rug, or even produce storage solutions that perfectly fit your space. By seeing objects you already own in a new light, you invite creative energy into your home, making the best possible use of nature's resources in your environment.

Recycle

Most items can be recycled in some way, but extra thought may be needed to determine how to best recycle them. Many items that can't be placed in household recycling can be taken to your nearest supermarkets, including soft plastics and electronics like phones and batteries. Stationery stores happily take used and discarded pens for recycling, bottle depots will give you money back for bottles and cans, and some clothing stores will take unwearable clothing materials for recycling into other textile products. There are so many options beyond your recycling bin that will give you some good brownie points with

Mother Earth, so it's worthwhile to look into them while tidying.

Discard as soon as possible!

Having a large pile of trash to discard makes for a hard time tidying up. When things get too big, recycle, or discard appropriately as soon as you can. This allows you to turn your attention to organising the things you love and neatening up your home, rather than spending space and energy to store the items you are getting rid of.

Once you have tidied and decluttered your space, determined that the items left are things you treasure or support your daily life, and are happy with the placement of items in your home, you can turn your attention to clearing and cleansing the energies of your home and yourself.

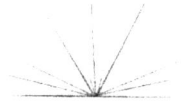

AS HYGGE WITCHES,
WE ACKNOWLEDGE THE SACRED IN THE
SIMPLE
WE MAKE THE MUNDANE MAGICAL
WE CREATE A HARMONIOUS ATMOSPHERE,
WITH FEELINGS OF WARMTH AND AN ENERGY OF
CONTENTMENT
WE CULTIVATE GOOD HABITS TO LIVE
COMFORTABLY
WE CONNECT TO OURSELVES, TO NATURE,
AND TO OUR HOME
WE PROTECT ALL THOSE THAT DWELL
WITHIN OUR SPACES
OUR HOME IS OUR SANCTUARY

Courtney Hope

SECURITY

Security and protection may not be the first things to come to mind when thinking of hygge, but they are important elements in practising Hygge Witchcraft. Our homes are our shelters; they should be the places that we feel most safe and held. Your home is a sanctuary, a place to wait out the storms, to celebrate the small moments, raise a family, fall in love, and play with your pets. The Hygge Witch's home is sacred and should be cared for as such.

But to the Hygge Witch, home is not just where you go to after a long day's work to put your woolly-socked feet up. It's not just a 'place'. Although houses are generally structures, a home 'has a spirit all its own, derived from the energy of the land it sits on and the vibrations of its natural materials [with] which it was built'. The architecture of the home we create is both physical and energetic. It is made up of the bonds of our relationships and habits, furnished with the objects and possessions that mean the most to us.

Building a shelter that makes us feel reassured and comforted is a basic foundation for the safety and peace of mind crucial to hygge. As a feeling of contentment, belonging, protection and connection, hygge shares the qualities that come with the idea of 'home', tied not necessarily to a physical structure, but rather to the experience of sanctuary in passing moments.

Hygge witchcraft connects this feeling of sanctuary with both your physical environment and your psychological peace. It moves daily routine and domesticity into interior alchemy.

Just like *KonMari's* method of 'waking up' the energy of items, the structure of your house itself wakes up energetically with the rhythms of your daily routines. Changing with the ebbs and flows of life, the natural energy in your home shifts, making cleansing, banishing, and protecting essential parts of the domestic routine of a home.

Because homes are a place of collective energies, sometimes unwanted energies get stuck in your home's actual fabric. Whether from previous tenants, people in your daily life throwing out toxic intentions, an energetic shift in visitors, or even a nasty bout of the flu, unwanted energies can cause the balance in your sanctuary to be off, bleeding discontentment into your overall feelings of happiness and hygge. It's important to pay attention in these moments and work towards wiping the energetic slate clean.

Thankfully, there are easy ways to do just that, which I list in my grimoire. Just as it's energetically important to physically declutter and clean your home, it's important to cleanse your space of unwanted presence. I've listed some suggestions, including brushing the energy away with besoms, saning, and cleansing (pg. 63).

Once you have energetically cleansed your home, you can get to work on providing protection, ensuring the negative energies you directed out of your home will not return to you. Methods of protection can include lining the windows and entryways with black salt (one of my favourite protection spells pg. 90), placing protective crystals such as obsidian and tourmaline at entrances and over doorframes, and placing lucky protective plants at your front door. These may seem like simple methods but projecting your desire for protection and safety by mixing it with these strong vibrational items will help keep your sanctuary sacred.

It's important to note that what you consider 'home' may look very different to another person's sanctuary. You may be lucky enough to own a house –a rustic hygge property on the outskirts of town or a neat suburban family home. You may rent an apartment in the city or share a busy and chaotic space with your family. Your sanctuary and your hygge may be found in your bedroom, where you get privacy and solitude after a busy day or driving in your car with the music blasting. Your hygge home can be the peace and quiet of your office, or maybe even outside of any solid structure as you adventure your way across the world with just the contents of the bag on your back.

Whatever it is, it doesn't matter what your 'home' looks like – it only matters that it's where your heart is.

Courtney Hope

CONNECTION

Humanity's heart lies in community and connection. Hygge tends to arise through developing and fostering relationships, falling in love, connecting with others, and enjoying little shared moments. Hygge is the soft moments; sharing a meal and a discussion with someone that you care about and revelling in the moment of cosy tranquillity that comes from it.

Connections can be formed through a variety of different means, and not with fellow humans. We develop deep bonds with animals and our pets, our material possessions, our surroundings, and even the house plants we struggle to keep alive. One of the first things that witches learn is how to ground to connect with this present moment on Earth. The practice of becoming present in your current situation will allow you to connect better with your intuition and the things your guides may be trying to tell you.

The best way to ground yourself is to sit with a short *meditation*, a 'mental exercise (such as concentration on one's breathing or repetition of a mantra) for the purpose of reaching a heightened level of spiritual awareness'

Although meditations can take many different forms, one of my favourites is to focus on your breath, inhaling and exhaling slowly to the count of four. As you begin to connect with the feeling of the ground beneath your feet, imagine yourself growing roots from the tips of your feet deep down into the earth, connecting you deep within its core. Send a burst of concentration deep into the world around you and sit with this feeling, letting thoughts pass by like clouds in the sky. By engaging with your breath and connecting your body to the earth, you ground yourself both energetically and physically.

Doing grounding work before every spell is regarded as essential in most witchcraft practices, but how long you do it for and how you do it should always rest on what feels right for you.

While you are in this moment of grounding you can start to connect to your inner voice, opening yourself up to the reason behind your spell work today, whether cleansing the space after performing a domestic chore, resetting your altar (pg. 75) on a full moon, or creating a spell jar to promote a happy home. (pg. 87)

By developing these connections to the moment, you can connect deeper into the things around you, from your home or garden to the wider world. Connection and mindfulness are like a muscle, best toned with exercise. By practicing connection, hygge witches develop a natural sensibility to stay grounded, enjoying and observing the present moment and her connection to it putting away distractions to focus on the things she loves and creating warmth, comfort, and belonging. Through connection, hygge is a state of pleasant wellbeing and security with a relaxed frame of mind and an open enjoyment of the immediate situation.

By developing her supernatural craft through this work in grounding and meditation, a Hygge Witch creates a strong and important connection with the natural elements of life.

Nature is the strongest force there is. Many spells and craftwork rely on natural items and power sourced from nature's deep energetic wells. Submitting your work to the universe provides a deep connection between you and the Earth; this is increased by enhancing your craft with natural objects, including items for spell work like moon water, a bowl of water that has been blessed by the light of the full moon for seven hours, blessings and remedies made from natural, and crystals derived from the body of the Earth herself. But there is more to connecting with nature than just the objects we use in our spell crafting.

Embracing the Earth in all aspects of your life is not only satisfyingly witchy, but also an important way to soak up the undeniable mood, health, and well-being boosting benefits of spending time with Mother Nature.

Courtney Hope

WELLNESS

The COVID pandemic that swept across the world made many changes in the way humans live, work, and function in society. Modern society's fast-paced environment had made rushing from activity to activity, stressing out, and constantly being plugged in and available the norm. When the pandemic hit, many people were confined to their homes to protect themselves and others, giving them the time and space to re-evaluate what was important to them. Long-ingrained behaviours and habits that were suddenly taken away, and some people began to find wellness and gratification through embracing a slower and more intimate way of life.

Wellness is a popular term that encompasses prioritising your self-interests and maintaining a healthy lifestyle of body, spirit, and mind. The term is multidimensional, focusing on holistic and natural approaches, self-healing, and preventive care . Many witches find their practice as part of a healing journey, using witchcraft for self-help and sustainable wellness rituals.

Often people think of *wellness* and picture expensive meditative yoga sessions, luxurious spa days, and long talks in a therapist's office. The 'Wellness Economy' is big business and accounts for roughly 5.6% of global economic output as recorded from the Global Wellness Institute in 2017 but there are many ways to practice wellness that don't require consumption, professional services, or expensive athleisure wear.

Wellness can be achieved simply by mindfully adjusting your daily habits. It includes nourishing a healthy body through nutrition, regular exercise and sleep, exploring the world through learning and problem-solving, and connecting and engaging with others in meaningful ways. For Hygge Witches, this may be harnessing positive energies to create your own magical tonics and beauty regimens, grounding in and engaging with nature, and conducting spell jars for health and positivity. Wellness looks different on everyone, but focusing on activities you find fulfilling raises your wellbeing and helps to provide you with a purpose in life.

This purpose affects your mental, emotional, and physical health and how you interact with the world around you. Engaging with your environment in a positive way by spending time in the garden, feeling the earth beneath your feet, and utilising the gifts of Mother Nature is a wonderful way to boost both your wellbeing and your witchcraft practices. Using this engagement in your spell work helps you reconnect to earth's big picture, as you start to recognise the cyclic nature of things and experience a deep awakening in environmentalism and sustainability.

Seeing the connection between living sustainably and a slow, comforting lifestyle is important to the hygge way of life. Hygge is all about making the time and effort for what is important, slowing down the pace to reconnect with the living world around you, creating moments of peace, joy, and mindfulness. It emphasizes the little things and of replaces your messiness with the mindfulness that sits at the heart of the hygge concept.

Living in a state of perpetual stress and worry plays havoc on your *sympathetic nervous system,* or **SNS,** responsible for your body's rapid involuntary response to a dangerous or stressful situation – commonly known as a "fight or flight" reaction.

When you feel stressed, your body shifts its energy away from functions not needed in a life or death situation like your reproductive cycle and sex drive, food digestion, and waste removal and toward releasing the stress hormones adrenalin and cortisol to prepare for survival. When the crisis is over, your body returns to its usual state with the help of the parasympathetic nervous system, or PNS.

Chronic stress can have long-term effects on the body, causing areas of your body to underperform when you are stressed, even if there is no actual danger. This can cause weight gain, long-term damage like hyperglycaemia and diabetes, bowel and kidney disease, heart palpitations and insomnia.

Living in a state of fast-paced stress takes a toll, but the social changes caused by the COVID-19 pandemic made it apparent that a slower way of living can lower unneeded stress, enhance wellbeing, and allow you to focus on your connection to the world.

My pursuit of hygge began with the concept of slow living as a mindfulness exercise. Once I was able to take stock of what was important to me, consume less, plan more, and take enjoyment from the small things, I was able to lower a lot of the unneeded stress on my body and could redirect my energy toward my connections and surroundings. By learning about witchcraft and using it as a modern wellness practice, I began to discover that the feeling of warmth and contentment in my surroundings I cultivated bled into other areas of my life.

I read multiple books on the subject of wellness, then used those beautifully designed books as hygge décor. As a practising interior designer, the elements of hygge design became important in my work and my personal life. I transformed my home with hygge comforts, filling my spaces with a plethora of blankets and pillows, my favourite books, and scented pumpkin spiced candles and spending time creating the perfect food and drink accompaniments (pg. 107) to enjoy the world I had created for myself.

My pursuit of hygge started out with design principles and objects like an adorable new throw blanket or rug, but the concept became so much bigger than just being happy in my home. It helped me re-shape my mind, re-examine my values, and create a simpler existence. Living the hygge way gave me more time to drink wine and read books and more positive feelings of comfort and happiness than I had ever had before.

COMFORT

If you Google the word *hygge*, you will be met with warm, comforting natural elements and materials like wicker, wools, and faux fur. You'll learn that the hygge aesthetic is having chunky knit blankets, warm faux fur throws and rugs, and comfy soft cushions layered together, and that using organic materials such as bamboo baskets are great to house the baggy food items in your pantry.

I love the hygge aesthetic. It is cosy, warm, and inviting. There are many different ways you can incorporate Hygge Witch living into your interior decorating. Hygge decorating is about what feels comfortable to you; some Hygge Witches may choose to make their rugged brick walls the feature point in their kitchens, while others may prefer to focus the space on their sophisticated dark mahogany floorboards.

Many publications focus on cultivating hygge through aesthetically pleasing and natural, but sparse and minimal interiors. These suggestions miss the spirit at the core of hygge. While minimalism and hygge have many similarities, the term *minimalism* conjures up images of white walls, modern furnishings, and clean sharp lines, beige, and empty homes without personality or comfort. Many minimalist spaces look sophisticated and modern, but lack the warmth of an actual home; a place where you cuddle up and read books on the couch, drop your snow-covered boots on your pile carpet, and dance around your kitchen with your partner as you hand-roll pasta with an unattractive appliance.

But what do you do when you don't have space to store something that is functional but not pretty? Hygge doesn't care about that! This is where minimalism and hygge collide - in both minimalist and hygge design, possessions are kept only if they are valued, they are useful, or they are needed. One of the biggest myths when it comes to minimalism is that you need to deprive yourself of items you love in order to achieve a blank space aesthetic. This is simply not true! Both design concepts, applied properly, allow you to own and use items, whether a vase or a painting that you value for its beauty, a throw blanket that you cuddle up under during the winter, or a mug from which you drink your latte every morning.

The concept of minimalism is about avoiding the unnecessary to cultivate simplicity, utility, and elegance. It encourages quality over quantity and choosing items that matter. Hygge, at its heart, follows these same principles. By focusing on less, you can slow down and spend time doing the enjoyable activities that promote contentment and wellness. While minimalism and hygge look very different, the peaceful environment curated by minimalism's focus on the useful and beautiful and the simple frivolousness that creates a welcoming environment in hygge culture, both encourage the art of simplicity and slow living.

Using the right objects can be important when casting spells, going about your daily activities, and setting the tone for your life, but it is best to do so in a place that harmonises your wants and desires together to bring you peace. Your home is your sacred space, so you need it to feel sacred to you as soon as you walk in the door.

Being a Hygge Witch is all about embracing the idea of living a slow, well-meaning, content life, adding magic into the everyday. Although hygge is not something that can be purchased, it is something that can be easily recognisable in interior design trends.

The use of natural elements mixed with items you personally value and enjoy that make a place hygge, but don't be afraid to look to others for inspiration - even searching for hygge on Pinterest can help you decide your own hygge aesthetic. Investing time and energy into creating your hygge space can help you find your niche as a witch, using your own ideas and inspiration to set up and decorate your altar and your home. I've included some of the interior decorating rules I've used in my home and career to help you on your journey.

Interior decorating rules

Interior decorators use specific elements and principles to achieve a complete look in the house, bringing balance and harmony to an environment and linking traditions and culture between living spaces. To do this successfully, interior designers must tick off every checkbox in the 'Design Elements' and 'Design Principles' lists. Design Elements are the 'building blocks' of the space, and the Design Principles are the 'tools' applied to the space as ideas are developed.

The Design Elements are as follows:

Space

This design element looks at how much space is available in the room. Focusing on how the space uses positive and negative space together to balance the room, like organising and storing items to the 80:20 rule. Positive space features furniture, fittings, artworks, or accessories, while negative space is the blank, open space around the object. Leaving intentional empty space provides room to breathe and is an important part of designing your perfect hygge den!

Line

Line defines the sense of direction of the room, leading the eye to feature objects that join together expressing movement and growth. Vertical lines can make a room seem taller and wide spaces narrower. Horizontal lines widen the narrow spaces and bring the eye level down. For a hygge-inspired room design, use horizontal lines to bring focus to plush furnishings used for comfort, and ensure that the colour is soft to avoid any highly-strung energy.

Shape

The shape of the room sets limitations and parameters that you have to work with, such as window shapes and styles, door types, and architectural points of interest. Curvilinear shapes are good to bring attention to soft or feminine energies, while rectangular shapes provide a harder edge that evokes minimalism. Natural and organic shapes provide a more grounded, naturalistic element to the room through their forms. I always suggest working with what you have in front of you; the more organic the shape and design, the better!

Colour

Beyond just being pleasing to the eye, colours require psychological consideration and can affect mood. Different colours are associated with different feelings and reactions, a tool commonly used in marketing, like the red and gold found in food chain advertising due to their close relationship to hunger. The colours that you use in your room while decorating affect the overall mood, so it's important to observe a basic understanding of colour. Lighter, more relaxing hues like creams, champagnes, whites, tan, and greys create an airy atmosphere in a room, while darker colours can bring a large room closer.

Texture

Texture is one of the most important elements in design, evoking sensory responses of touch. Different textures and surfaces add dimension to a room, with leather couches, thick book covers, and woven wool blankets giving a completely different feeling from cool linens, soft upholstered furnishings, and light wood. This is one of the most important elements of design when creating your hygge haven; the more organic and comforting the texture, the better!

Light

Light is essential to any interior space and can be introduced through the use of either natural or artificial light sources. Natural light from the sun, brought into the room through doors and windows, has different qualities from the wide range of colours, textures, and brightness available in artificial light. Use soft, warm, ambient lighting as from multiple comforting elements when setting up your home rather than direct, cool, bright white lights, as they can be harsh in the setting of a home. Explore multiple aspects of lighting in your space, combining lamps with industrial-style naked bulbs and heaps of fairy lights to ensure you never have to use the big ceiling light if you can help it!

Pattern

The 'repetition of a graphic motif', patterns can provide an added visual interest to any interior decorating. Ranging from the culturally respectful display of intricate blankets and other local Indigenous artworks to the natural grain of wood panels in your dining table or even just a group of three decoration items in asymmetrical sizes, there is a vast range of patterns to choose from for your home. Repetition is key for pattern design elements, so you don't want to clash too many patterns together unless you are working towards a maximalist style.

The 'tools' that are applied to a space being redesigned are known as 'Design Principles'. These are as follows:

Balance

In design, balance is the idea that objects have been placed with consideration to provide a sense of stability within the room. It is the arrangement of the positive and negative space to harmonise a room's design. Symmetrical balances provide the sense of equality and composition, while asymmetrical designs suggest growth and interest.

Rhythm and Repetition

Rhythm and repetition are very similar to pattern, as they provide visual movement across the space, through repetition, alternation, and progression. This can be achieved by putting three items together for a sense of symmetry, setting items out in a row or in a way that slowly gets bigger in size and shape. One of my favourite rhythm and repetition decoration ideas uses pumpkins as décor around Halloween. A collection of three medium to large pumpkins sitting on your front porch steps brings instant symmetry and Autumn energy to welcome you in your trick or treating.

Emphasis

Emphasis refers to a singular focus point that is very clearly the centre of attention, like the stunningly restored wood fireplace in your library, an artwork of your deity, or a low hanging chandelier. All visual roads lead back to this focal point, and the more natural the material of your focal point, the more positive the energy that will be taking up your space.

Proportion and Scale

There is a general consensus among designers that proportion and scale are important tools for creating balance and harmony in a room. The golden ratio is usually set at roughly 60:40, broken down to 60:30:10. This ratio states that you need to decorate 60% of a room with a dominant colour, 30% of a room with a secondary colour, and use the remaining 10% of colour as accents. For a hygge-inspired Scandinavian room, this could be 60% calming beige, 30% light-coloured woodgrain textures, and 10% pops of colour like blush pink or greyish blue couch cushions to add visual interest. Following this rule is guaranteed to create harmony and scale within any interior you decorate.

Harmony

Adding all these specific Design Elements and Design Principles together creates a sense of harmony, essential to the success of the overall design. Hygge inspired interior design should always find harmony in how well your favourite, comforting elements fit together. If you feel content, happy and harmonious in your space, you have reached the level of hygge you are trying to achieve!

When applying the principles of interior decorating to your home, some of the most useful advice is about ensuring a focal point. Choose a particular item – whether that be a lounge, a painting hanging on the wall, a painted over brick wall itself, or a piece of furniture – and decide that will be the room's focal point - the first thing to draw the eye when you walk into the room. Building from there, creating other small arrangements around the room that will draw the eye once it has taken in the hero piece.

If your focal point is a certain colour or texture, try adding hints of that colour around the room for a cohesive look, like couch cushions, a lampshade design, or even a table runner.

Another good tip is to utilise and embrace blank spaces. This is where the concept of minimalism and hygge can merge, as filling your space with a mish-mosh of random items can cause your home to look messy and cluttered rather than comfortable. Too much positive space can have a negative effect on the space's energy, so make sure items you place have plenty of space to breathe.

Decorating with the five senses:

Humans have five senses: sight, hearing, smell, taste, and touch. When decorating a home, I like to make sure all five senses have been appealed to on a subconscious level.

Decorating your home and ensuring it is neat and well-kept appeals to the sense of sight through the general rules of aesthetics and interior decorating in this chapter.

Although not often considered, the sounds you can hear when you enter a home impact how you feel. Loud traffic from the city below permeating your home can create a sense of ill-ease and anxiety; this can be mitigated through pleasant music and soft furnishings that dampen acoustic reverberation.

Creating a hygge smell in your home can come in many forms: beautiful smelling candles, incense, or even the scents of baking bread or cooking food wafting through your house. I am a personal lover of the smell of sea salt and caramel, which adds to the warmth of the home.

Decorating according to taste might not be constant in your home, but ensuring you have delicious food available in the form of snacks, a dinner party, or even just in jars in the pantry can add elements of hygge to a home.

Touch is an excellent way to add comfort to your home, with soft furnishings like faux fur rugs, chunky knit blankets, and other textured items providing pleasant textures to touch and feel.

Decorating with all five senses helps you establish the perfect balance in your home, making sure both guests and residents have everything they need to be comfortable.

RITUAL

Being a Hygge Witch is all about finding the magic in the mundane and turning everyday activities –doing the laundry, cooking your dinner for the evening, and working in your office – into magical moments to be present in. Your daily routine is ritualistic, whether you are in the daily grind or not, so it is beneficial to treat each routine item on your 'to-do' list as its own intentional ritual. There is power and spirituality in daily rituals, from getting up and doing meditations every morning to pulling a tarot card for the day. Rituals keep your manifestations in the front of your mind while you work for them.

Hygge Witch Rituals

Here are ten short and slow rituals that you can do to add a bit of Hygge Witchcraft into your everyday life.

1. **Yoga** – Whether you are performing salutations to the sunrise or loosening tight muscles, a morning yoga and meditation session can do wonders for both the mind and body, slowing you down to take in the moment in its entirety.

2. **Meditation** - Meditation can improve many of the body's functions, such as breathing, blood pressure,

and the parasympathetic nervous system, but it also helps us actively learn how to transform our thoughts.

3. **Journal Before Bed** – A private journal is a wonderful place to collect your thoughts and reflect on the day you just had. Remember to always finish your entry off with three things you are grateful for, no matter how big or how small. I have found my perspective becoming more positive just by finding the gratitude in everyday moments.

4. **One Card Tarot Pulls** – Every morning, shuffle a Tarot card deck and pull one of the cards with intention related to your day. Keep this tarot card in mind as you go about your day, reflecting once on it in your evening journal.

5. **Use Essential Oils** – Using essential oils can provide beautiful aromas to any home, but you can also use specific infused essential oils to enhance domestic chores, like adding them to the solution you use to wash your floors (pg. 72), freshening laundry with juniper berry, or spraying lavender onto your bed before sleep.

6. **Give Yourself a 'Gua Sha' Facial** – This traditional Chinese medicine practice has had a huge resurgence in popularity due to its ability to firm the skin of the face and neck, relieve muscle tension and help with lymphatic drainage in the face. Done with a variety of shaped crystals and rollers, you can choose a crystal for your Gua Sha facial to focus your intentions, like rose quartz for self-love.

7. **Cook with Intention** - Light a colour coordinated candle (pg. 77) and set an overall goal you wish to meet in the provision of this dinner, such as 'to nourish every

person at the table'. Use ingredients that supercharge your intention and cook with love to mix it all together.

8. **Go for a Walk** – People that have dogs have experienced the joy of this simple act, because even on our darkest days pets pull us into the world. Whether you have a pet or not, a walk can be a beneficial chance to disconnect from electronic devices and take in the world around you, even if just for a minute. Appreciate one thing and see how this connects you to your present moment.

9. **Create Your Own Beauty Items** - Using natural ingredients, it is quick and simple to make your own body scrub with sugar, coconut oil, and rose scented essential oil to promote self-love or a soothing rose-hip facial oil infused with full-moon charged amethysts.

10. **Light Candles** – Used to signify different properties, the colour and scent of candles can be used to support your intentions and enhance your other rituals.

While Hygge Witchcraft tends to focus on the domesticity of daily life and ritual, there are also auspicious times for conducting certain spells, usually associated with the lunar calendar. It is good to familiarise yourself with your local lunar cycle and any events that are regularly celebrated by witches.

It can be confusing to coordinate celebrations across the equator, so focus solely on the events in your hemisphere, with a general rule in the southern hemisphere that the magical musing of the date will be the opposite of European tradition. Working with the atmosphere around you is more important than working the specific dates set, so tune in to what is going on in your area rather than what the celebrations are like somewhere else in the world. This connects you more deeply to your current surroundings and the nature under your feet.

MOON PHASES

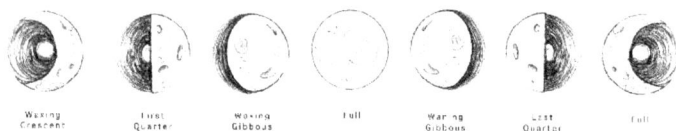

| Waxing Crescent | First Quarter | Waxing Gibbous | Full | Waning Gibbous | Last Quarter | Full |

Moonology

Getting into sync with the Moon is one of the most basic witchcraft practices and one of the first things people tend to take an interest in when learning the basics. *Moonology* is a specialised form of magic, in the same way that candle magic, crystal magic, sex magic, and even divination are, and it is possible to go as in-depth as you would like by following your moon and sun signs into their planetary houses. Working with the basics of Moonology, like the regular rotation of the moon, is one of the best ways to get in sync with your witchy calendar, allowing you to plan when to conduct spells. By conducting spells on their most auspicious dates, you make the intention behind them more powerful and thus more likely to occur.

The lunar cycle sees the new moon and the full moon occurring every two weeks in rotation. I recommend looking up the Moon's position in your area online and jotting the moon cycle in your calendar or your phone's diary so you can work accordingly.

The Moon's cycles are:

- **New Moon** – This is the time to plant the seeds of your future dreams. Indulge yourself with a massage, hot sex, or a hot bath to help make you feel alive.

- **Waxing Moon** – This occurs three and a half days after the new moon and lasts three days. This is a good time to move forward and explore your dreams.

- **First Quarter** – This takes place seven days after the new moon and is a good time to commit to your goals as you move forward.

- **Gibbous Moon -** This moon occurs ten days after the new moon, and is a good time to gain momentum in your spell work.

- **Full Moon** – This moon takes place fifteen days after the new moon and is one of the most important times for spell work. I usually conduct Coven meetings every full moon, as this is a great time to set intentions and bring your work to fruition. Don't forget to put your crystals in the full moon's light to charge!

- **Waning Gibbous** – Taking place three and a half days after the full moon, the waning gibbous is a good time to look retrospectively and work towards your goals. Do some breath work during this moon.

- **Last Quarter -** Seven days after the full moon, this moon is the time to re-evaluate and allow your goals to take shape.

- **Waning Crescent** – Ten and a half days after the full moon, this is the perfect time to surrender and rest after your hard work.

The moon takes 28 days to go through all twelve signs of the Zodiac, spending just over two days in each sign. This means that as the moon travels through the star signs, there will be a new moon and a full moon in each sign. These moons mean different things depending on the Zodiac sign.

Many sources focus on the different meanings created as the moon moves through the sun signs and the moon signs. Different magical items like crystals and candles can also interact with the zodiac signs and the moon's movements to enhance different energies and intentions.

If you are interested in coordinating your spells to follow the zodiac signs, it will mean more in-depth and constant spell work. Because of this, I suggest working closely with the basic lunar cycle until you feel called to investigate further.

The Witches Holidays

Just like other religious and social communities, Wiccan practitioners observe special holidays throughout the year, called the Sabbats. Modern Pagans celebrate the Sabbats in accordance with the movements of the Sun and the seasons.

These holidays are usually related to Celtic agriculture festivals, with traditions related to harvesting and planting at that time of year. Many celebrations include a ceremonial bonfire, linked back to ancient Viking rituals and a key feature of many festivals throughout history, but there are plenty of other ways to celebrate if you can't have a bonfire.

Samhain / Halloween

Samhain is the witch's new year and takes place on October 31st, which is regularly celebrated as Halloween and the Day of the Dead in the northern hemisphere. Samhain is when the veil between worlds is the thinnest, and there are many beautiful celebrations and rituals that can be conducted at this time. For traditions with Celtic and Nordic roots, November 1st is the most auspicious day for this celebration, usually characterised by lighting a candle for a loved one and placing it in the window.

In the southern hemisphere, Samhain technically falls around 1 May, in mid-autumn. As the veil between worlds is thin, this is a perfect time to honour deceased loved ones by visiting their graves to clean them up and delivering offerings of candy, flowers, money and other colourful objects. I also like to use this time to conduct tarot card readings in connection with my dead ancestors, as it is an auspicious day to be led and informed by spirits from beyond the mortal realm.

Samhain Tarot Reading

1. Me in the moment.
2. Ghosts of the past.
3. Spirit of the future.
4. Behind the veil.
5. In front of the veil.

Yule/ Winter Solstice

The Winter Solstice, or Yule, is celebrated at the height of winter - from 20–23 of December in the Northern hemisphere, sharing many of its traditions with Christmas. In the southern hemisphere, the celebrations take place between 20–23 June instead. Regardless of which month you celebrate it, Yule represents the transition of the seasons and – on the solstice - is the longest night of the year. Yule is a time to have a winter feast with loved ones, enjoying mulled wine and other festive delicacies in front of the fire (pg. 107).

One of the primary traditions shared between Christmas and Yule is the 'Yule log' or 'Yuletide log', burned in the fireplace during the 12 Days of Christmas to ward off evil spirits in your home. If you don't have a fireplace in your home, or if you're celebrating in June, you can switch out the burning of the Yule log for an edible Yule log cake instead, which adorns the table for twelve days before being eaten at the feast.

Imbolc

Imbolc is the Pagan festival that welcomes the signs of new spring life to fields and pastures. Celebrated on 1st of February in the northern hemisphere and 1st of August in the southern hemisphere, it is the first of three Pagan fertility festivals. In addition to the giant bonfire, lit to drive away the winter chill, witches also burn seven candles on their altars over seven days as a way of re-dedicating themselves to their practice.

The main Goddess celebrated during Imbolc is the Irish maiden Brigid, the Celtic goddess of fire, healing, fertility and smith craft, so celebrating witches create straw Brigid's Crosses or corn dolls as talismans for fertility and the new season. Associated with fertility and love, Imbolc celebrations can also introduce traditions from Lupercalia, the witch's version of Valentine's Day.

In Nordic countries, children dress up as witches for an Imbolc carnival celebration, playing the traditional game of 'beat the cat off the barrel'. Originally played by hitting a barrel repeatedly with a stick to get a real live cat out of it, the cat is now symbolised with a picture of a black cat instead of an actual cat and the barrel is filled with candy or oranges similar to a birthday party piñata.

Ostara / Spring Equinox

Ostara is the spring equinox. Usually celebrated between the 20–23 of March in the northern hemisphere and the 21–23 of September in the southern hemisphere, is a time when night and day are of equal length and serves as a reminder to both honour the past, celebrating what has been achieved during the winter, and to plan for the future, planting physical and metaphorical seeds to nurture through the upcoming year.

This is the time that you should be conducting a lot of your gardening and green witch activities, like planting trees and herbs and making flower crowns for your altars. In the northern hemisphere, Ostara often coincides with Easter, so it is not unusual to find rabbit icons, eggs, and chocolate perfect for celebrating at this time.

In the Nordic countries, birch branches and coloured ribbons are used to make besom bundles. Children used to beat their parents with these bundles on the first day of spring to symbolise rebirth and fertility. Now there are fewer beatings, with witch hazel besom bundles used as aesthetic wreaths instead, but the association is still the same.

Beltane

Beltane is usually celebrated on the 1 of May in the northern hemisphere and 31 of October in the southern hemisphere, so is often referred to as May Day. Many traditions used to celebrate Ostara can also be utilised for Beltane, as both festivals represent similar ideas of fertility and spring, but a tradition unique to Beltane is the Celtic May Pole, a large upright pole decorated with ribbons that dancers use to create beautiful patterns and shapes.

Litha / Summer Solstice / Midsummer

The Summer Solstice, or Litha, is usually celebrated from 20th-23 of June in the northern hemisphere, and 20th-23 December in the southern hemisphere. Litha is Midsummer, the longest day of the year and the shortest night. In Denmark, Midsummer is celebrated on St. John's birthday, the 23 of June, by burning a fabricated witch on top of a bonfire in remembrance of the people who were persecuted during the 16th and 17th century's wars of religion. It's probably not the most festive part of the celebration for witches in Denmark to see an effigy of themselves being burnt at the stake, but this tradition maintains its popularity as a ritual for protection against evil forces.

To celebrate without burning an effigy, you can have a summer bonfire or BBQ, walking three times around the fire to promote wellness all year round. Midsummer is a joyous time filled with abundance, and can also be celebrated with a drum circle, a beautiful walk-in nature, or -for those in the United Kingdom- a visit to Stonehenge, which was designed specifically to celebrate Litha.

Lammas

Lammas, sometimes referred to as Lughnasadh, is a harvest festival that usually falls on 1st of August or the 31st of February, depending on your hemisphere. Marking the start of harvesting foods and storing them for winter, Lammas is celebrated by baking bread from scratch, making beeswax candles, and blackberry picking.

Mabon / Autumn Equinox

The Autumn Equinox is another harvest festival, usually celebrated between 20th-23 of September in the northern hemisphere and 21st-23 of March in the southern hemisphere. Associated in America with the tradition of Thanksgiving, the Autumn Equinox is a time to give thanks to the spirits, traditional practices, and the Earth for providing food for winter. To celebrate Mabon, have a working bee with your fellow witches, cultivating your herbs, fruit, and vegetables in your garden, creating oils, tinctures, and jams out of your harvest, and hosting a delicious feast.

Using these festival periods to conduct specific spells, charge specific crystals, or set particular intentions can help you connect to tradition and community but being a Hygge Witch is also about incorporating magic into your daily practice. Follow your instincts and your intuition; if you feel like you need to create a spell candle or do some grounding work, go ahead and do it, even if it doesn't necessarily align with the celebration or the moon phase.

Practicing witchcraft is all about doing what is right for you when you need it. Above all, make sure you follow your intuition to give yourself what you feel you need right now. This way, you will never put a foot wrong.

GRIMOIRE

Cleansing Magic

Cleansing Your Home with Smoke

Cleansing the house of unwanted energies is a great way to hit the reset button on your home. Extremely charged energies can leave behind a residue that can be felt throughout the years to come. Performing a blessing on your space releases the negative energies out into the universe and sets an intention of protection into your home.

There are many ways you can perform a cleansing or blessing ritual, but whatever way you choose must speak to you. If you struggle to find the words for a cleansing ritual, I recommend you meditate and connect with your breathing (pg. 33). As you meditate, the words will come; even if they sound silly in your head, there is a reason that this particular string of words is running through your mind. Think them with intent.

Whenever I bless my house, I light my chosen herb bundle and travel around the room slowly, letting the smoke drift as deeply into the corners of the room as it can. If smoke is lighter in one area, fewer negative energies have permeated there. As I float my chosen herbal stick around the room, I think to myself something along the lines of, *"Oh great universe, I submit unto you these words of blessing, asking you to protect my home and all that dwell there."* Ensure that you can really feel the positive energies you are bringing into your home. Once finished with the herbal stick, set it aside to stop burning.

When choosing a tool to cleanse your home, be careful to ensure you don't appropriate closed spiritual practices. Many Indigenous cleansing practices have made their way into the mainstream, like the white sage and palo santo, despite their significant spiritual importance to Indigenous communities. To avoid cultural appropriation from closed spiritual practices, I suggest you utilise more natural 'saning' items from your local environment. In Australia, this includes bush herbs and Eucalyptus, while in Danish peasant tradition wormwood is commonly used.

Each material has a fragrance with its own intentions and reasoning for use in spells, depending on what you are looking to purify and what energy you would like to invite:

- **Sage –** Used for the cleansing of people and places.
- **Palo Santo –** Used for peace, harmony, health and luck.
- **Rosemary –** Used for clarity and motivation.
- **Juniper –** Used to reset energies and provide stress relief.
- **Yarra Santa –** Used for self-love and growth.
- **Cedar –** Used for confidence and strength.
- **Eucalyptus -** Used for protection and healing.
- **Sweetgrass -** Used to encourage positive spirits to take up residence.

How to Make Cleansing Spray

Cleansing spray helps to clear the negativity from the area and is a good substitute for burning herbal sticks in the home, especially if you aren't able to burn substances indoors. To create cleansing spray you will need the following ingredients:

- Filtered water
- Clear grain alcohol like vodka or gin
- Essential oil) use list above to choose which kind)
- A mixing bowl and a glass spray bottle

To make the spray, first mix two parts of water to one part alcohol and one part sage essential oil in the mixing bowl. You can choose to put in crystals such as clear onyx or quartz here to help supercharge the mixture. Set the bowl out overnight under a full moon. The next day, decant the liquid into your spray bottle and cleanse until your heart's content!

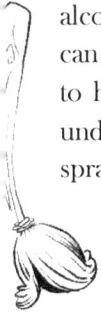

Cleansing Your Home with Brooms

Witches are sometimes depicted with a pointy hat riding a broomstick across a full moon, a caricature with a long tradition. One of the first references to witches riding brooms was in 1453, when a clergyman named Guillaume Edelin confessed under torture to making a deal with the devil and flying to the Sabbath on his broomstick. Around this time rumours began to circulate that witches would use brooms as a kind of sex toy, slathering them in hallucinogenic ointment on it made from ergot, a type of rye fungus with LSD-like qualities. Some historians have even suggested ergot was responsible for mass hysteria events, including witchcraft panics, and in 1976

Dr Linnda Caporael of the Rensselaer Polytechnic Institute put forward the fungus as the real cause of bewitchment in the Salem Witch Trials .

Witches have been known to use brooms or besoms, for magical purposes like cleansing the home and sweeping energy through the house. The besom symbolises cleansing negative energies and is used by 'sweeping' the floor a few centimetres above it so that the bristles are not actually touching the floor. Besoms can also be used to contain a magical circle or create a doorway.

Used in ritual cleansing and some hand-fasting ceremonies, besoms can either be purchased or made by the witch themselves out of tree twigs, twine, and a large stick. To make your own besom, simply bundle the twigs together until they are an appropriate length for your besom and place the end of the large stick through the middle of the bundle, securing the twigs and stick together with twine. Once you have energetically swept the room; you can leave the besom bristle side down by your front door as an extra protection charm.

Cleansing Your Home with Sound

I love attending sound healings and using singing bowls, but did you know that you can also energetically cleanse your home with sound? Whether you clap, chant, sing, play hand drums, singing bells, gongs, and wind chimes, or use a tuning fork, sharp, sudden sounds - often metallic - vibrate on a different frequency undetectable to human ears. The sound compels objects and the atmosphere to resonate in harmony, neutralising the space and penetrating resistant energy blocks.

Similar to smudging or saining, where the smoke is more concentrated in areas with stronger negative energy, the duller the noise of a sound cleansing, the denser and more clogged the energy in that area is. It is important to note that the sounds used in cleansing must be pleasant to provide the right balance of energy in the room, and that you should always open the windows when clearing the space so that the energy has a way to depart.

Cleansing Your Home with Crystals

Crystals can also provide cleansing qualities for your space. One of the objects most closely associated with modern witches, these gemstones are more than just pretty rocks. Crystals are used in alternative healing to harness the magic within the mineral's structure and send its energy as an intention out into the universe.

Different crystals have different uses in magic and can be powerful components of spell work. Before you use crystals to cleanse your home, it is recommended to cleanse, ensuring you thank the crystals for drawing negative energies away and sending them into the universe.

There are many different ways you can cleanse crystals depending on the type of crystal you have and the element it aligns with.:

- Cleanse Earth elemental crystals (amber, garnet, ruby, obsidian, black tourmaline) by burying them overnight in the dirt of the Earth.
- Cleanse crystals associated with the Wind element (amethyst, moonstones, selenite, rose quartz and

fluorite) by smudging and running the crystal through the smoke.

- Cleanse crystals associated with the element of Water (aquamarine, emerald, green aventurine, and blue lace agate) by passing them under moving water, either running cool tap water over them or placing them securely at the bottom of a nearby running creek (but be careful not to lose them!)
- Cleanse crystals aligned with the element of Fire (citrine, malachite, carnelian and red aventurine) by smudging or running the crystals through a candle flame.

Although all these methods are effective, my favourite way to cleanse crystals is a universal method that works regardless of element. I prefer to quickly bless my crystals, leaving them out overnight on the night of the full moon to charge.

Once you have charged your crystals, it is time to place them in the locations that best align with their meanings. Some of the crystals that are perfect around the house include:

- **Obsidian -** A jet-black crystal that cleans up emotional debris from the past and provides protection from addiction, fear, anxiety, and anger. Place a charged obsidian crystal at each entrance of your house to keep these negative energies out...
- **Black Tourmaline -** Another jet-black crystal that grounds and protects. Leave this crystal by the entrances of your house to provide protection and grounding.
- **Rose Quartz -** This light pink crystal is associated with love and can be used to resolve anger and bring

harmony to high-running emotions. It can also power self-love if you surround your bed with charged rose quartz crystals.

- **Amethyst** – One of the most popular crystals in use today, this luxurious purple gemstone can assist with familial bonding and creating a calm, stress-free environment. Amethyst works particularly well when situated in the lounge or family room, clustered together with gems representing each member of the household.

- **Clear Quartz** – This clear crystal is prized for its abilities to both clear the mind of negativity and enhance the power of other crystals, reflecting their healing vibrations. Use clear quartz crystals near any other crystal for a boost in their energy.

- **Citrine** – These are great crystals for the mind, blocking any negative energy that might be affecting your concentration. Use citrine crystals on your desk, in your office, or anywhere else you need to do serious work.

- **Carnelian** – This fiery orange jewel is perfect for boosting creativity, enhancing motivation, providing extra confidence. It's the perfect crystal to use in the kitchen, sitting on a windowpane.

- **Selenite** - This calming moonlight-coloured crystal has incredible relaxation powers. It's also a catch-all crystal that can be used to charge other crystals near it! Selenites are often made into lamps, which provide beautiful ambiance to any home while enhancing its vibrations.

Make sure to regularly cleanse your house and charge your crystals on each full moon, ensuring that the negative energies they have trapped are consistently released out of your home the crystals have had their energies reset for any future work you do with them.

Cleansing Your Bed

Although it isn't a typical cleansing spell, I believe this bed enchantment that will always ensure good sex is a home must-have.

I love having rose quartz crystals surrounding my bed when I sleep to provide an influx of self-love, but I also like being a welcoming host. To cleanse negative energy from your bed, imbuing it with positive frequencies and romantic intentions, all you need is a sprinkling of rosewater and some rose petals.

Sprinkle small amounts of rosewater onto your bed, followed by the rose petals. Next roll around on the bed until you start to laugh, focusing on your intentions as you ask the universe to bless your bed and the actions in it, ensuring good and fulfilling sex at all times.

Protecting Your Home with Black Salt

Salt is a cleansing and purifying agent. It is commonly used outside of witchcraft, especially in folklore, like the phrase *"an unbroken ring of salt will protect you"*. Living with a Himalayan salt lamp (or two) will not only amplify the hygge factor in your home by introducing natural elements and textures, but will also clear the air, soothe allergies, and promote calmness.

If you want some further protection of your home, there is an old folk mixture called *black salt* that is extra powerful for banishing, protecting, and cleansing. Black salt can also help protect against the effects of the dreaded Mercury Retrograde that wreaks havoc on our lives, making it one of my favourite ways to protect my home.

To make black salt you will need the following ingredients:

- Sea salt
- Activated charcoal
- Protective herbs such as lavender, rosemary, or sage
- Black pepper
- Earth from the garden
- A mortar and pestle

Grind and mix all the ingredients together in the mortar and pestle to a fine powder. Seal the powder inside a witch's bottle for preserving, or sprinkle in an unbroken line across your doorways and windows of your home.

This will protect negative banishing or hexing energies from entering your home until the line begins to break. Bury the remains of the black salt in the garden bed once the ritual is complete, or until you are ready to draw another unbroken line.

How to Make Rosewater

Rosewater is used in spells as a symbol for good luck in love and passion and can even be worn as a perfume if you're looking to really send your date wild.

What you need to create rosewater, you need the following:

- 6 Roses of the appropriate colour. Use red for romance and love spells, pink for friendship, and white for peace.
- 2lts of water

Heat the water on medium. While the water is bubbling away, peel off each petal from the rose and let it fall in the water, focusing on your intention. Once all the petals are in the water and it has returned to simmering, lower the heat and cover for half an hour. Turn off the stove and let the mixture cool before you remove the petals and bury them. Drain the remaining liquid into a witch's bottle after diluting the mixture with 2 litres of cool water.

Protective Floor Wash

It may seem like a strange suggestion, but if you are washing your floors anyway (and you should be!) why not charge up your chore with a little bit of extra hygge magic? This floor wash recipe comes from the amazing *HausWitch* Erica Feldman. The ingredients you need are as follows:

- 10 drops of fir essential oil
- 10 drops of juniper berry essential oil
- 5 drops of orange essential oil

- 13 drops of tree agate essences
- 85 grams of concentrated hard surface cleaner
- 30 mls of vodka

Mix all ingredients together in a warm bucket of water, using the mixture to wash your floors. Doing this on a Sunday will create a perfect energetic start to your week, bringing balance, stability, and grounding from the floor up.

Cleansing your home is an important ritual, inviting positive energies into your life and setting your home's tone and intention. The positive energy of a physically and energetically clean home's clear intention adds to your hygge, creating a comfortableness that can be felt throughout the environment. Remember, your home is your sanctuary: you can set any intention you want, either out loud or in your head, because as long as you are comfortable there is no wrong way to set the tone of your home.

I recommend cleansing your home at times of change and new beginnings like the start of the month or a new season, after a particularly negative experience in the house, or whenever you feel that the house's vibrations need a good clearing. Working with the charge of the full moon is always the most auspicious, so make sure you check your local moon charts to schedule your cleansing rituals.

Courtney Hope

Decorating Magic

When it comes to what you consider decorative, the only rule is that there are no rules. Many magical items are beautiful enough to incorporate into your décor scheme. Here are some of my favourite ways to bring your magical work into the aesthetic of your home.

Creating an Altar

Every Hygge Witch's house looks different, but there is one staple furnishing that most have in common – an altar.

No longer just a sacrificial table covered in ram's blood or an ornate church frontispiece, the modern witch's altar is a place for her to practice her spells and focus her intent that is uniquely her own. Described as *"like a mood board come to life"*, your altar is a place to display the objects you use in your mindful witchcraft practices.

There are as many ways to set up and use an altar. Many redecorate according to the seasons, letting the blessings for each time of the year be their altar's strongest feature. Some use the altar to welcome the full and new moon, following the lunar cycle to meet their intentions. Others use their altars to activate their vision of the future, complete with photos and posters of where they want their life journey to lead, creating a vision board on which to meditate.

I've seen many different altars in my life, each one reflective of the many different witches I know, but the altar I created in my home focuses on the comforting, the content, and the hygge.

The first thing you need to do is find a place for your altar. Where you put your altar is important: it needs to be in a place where you can check it regularly to set your intentions and remind you of your touchstones. I originally placed mine beside my bed, but I found my magical workings made it hard to sleep. I realised that a lot of my spell casting is actually done in my lounge, facing north at the front door, so I created a small space on the bookshelf in this area for my work.

Once you have found a space for your altar, you need to cleanse and bless the space. I used herbal sticks and then cleaned the surface of the altar with Florida water or blessing colognes. Once the space had been blessed, I placed a large round cutting of wood on the shelf as the base for my altar, bringing in the natural hygge elements of living.

Next, all the objects you need for your spell work go on your altar. You might have a fixed candle, a spell jar, a votive candle or a chime candle, flowers and plants, and other natural or magical items. Ensure you select the items that go on your altar with care; your altar should reflect the beauty and intention of your spell work and be in line with the Hygge Witch vibes you are trying to create.

Once you have created your altar and set your intentions for the spell you are doing, it's important to ensure you maintain it with the same care you put into creating it. As a practicing witch you will be using the altar over and over again, repeating the cleansing and blessing processes for each spell so it remains in motion. You need to make sure that any items you won't use again, like representations of your intentions, wax, candles, and flowers used up by the spell, are disposed of in the appropriate way every time.

Candle Magic

Candles serve a functional purpose in magic spells as well as being decorative and aesthetically pleasing and are an excellent addition to your altar. Candles come in all different shapes, sizes, fragrances, styles, and containers, and nothing evokes the hygge aesthetic more.

When I think of hygge, I picture glowing candlelight, the flames burning dimmer and dimmer as the wax forms natural droplets on my altar. I'm not alone in this association - when asked what they associate most with the idea of hygge, 85% of Danes mention candles! Apparently 28% light candles up every day, and 31% always light more than five at a time .

The relationship between candles and hygge is especially good for Hygge Witches, because candle divination is one of the most popular forms of spell craft with modern witches. Candles represent all the elements together: the wick represents the Earth, the wax represents water, the oxygen keeping the flame alive represents air, and fire is the flame itself. This is why candle magic and its associations are extremely powerful.

There are many ways that you can incorporate candles into your work. You can enhance all your spell work by anointing the spell candle with oil and herbs, or even by using a significant object to scratch sigils into the wax before you burn it, as long as it all corresponds with your final intention.

Candle Magic Colour Meanings

The colour of the candles you use in your spell work symbolises different elements of life. By carefully selecting your candle colours even if you're using a fixed Deity candle, burning chime candles, or simply decorating your home with votive or taper candles, you can set your intention to its fullest extent.

The different coloured candles represent the following:

- **Red –** Symbolises health, strength, lust, sex, and independence.
- **Orange** – Symbolises creativity, inspiration, opportunity and will power.
- **Yellow** – Symbolises intellect, studying, confidence and memory boosting.
- **Green** – Symbolises finances, money, prosperity, luck, and abundance.
- **Blue** – Symbolises healing, tranquillity, happiness, and dreams.
- **Purple** – Symbolises royalty, communication, divine power, and psychic energy.
- **Pink** – Symbolises love, honour, compassion, children, and nurturing.
- **White** – Symbolises purity, truth, cleansing, protection, and blessings.
- **Grey** – Symbolises perfect balance and neutrality.
- **Black** – Symbolises binding, reversing, banishing, and absorbing negativity.

Fixed Candle Magic

Fixed Candle Magic is a very specific type of candle magic spell that uses candles sold for spell craft with the intention already fixed by the candle maker. As botanicas and conjure shops often sell these candles, so this is a great opportunity for you to support other witches, their shops, and their cultures. The most popular kind of intentions that you can find fixed candles for include:

- Money drawing
- Crown of success
- Road openers
- Block busters
- Good luck
- Health and healing
- Protection
- Love drawing
- Better business
- Better sex
- Cleo May

Regardless of your intention or the deity you are including in your practice, the spell work for candle magic is always the same. To begin, first bless your altar and your space, grounding as you usually would when conducting a spell. Place the candle on your altar and start thinking about manifesting your intentions. Write those intentions on a piece of parchment, folding your parchment in half three times facing towards you. Place your parchment on the altar and place any other offerings you desire to be used in the spell. These can include crystals, a glass of water or wine, or any other items that call to you.

Try to keep your candle burning for as long as possible, preferably until it fully burns out. Seven-day candles have a longer lifespan. Although it is expected that you burn your candle fully, you should never leave a burning candle unattended.

When you are finished with your spell, dispose of the candle in hallowed ground like burying it in a cemetery (respectfully and with spirit permission) or bless and recycle the candle, setting the energy from the spell free into the universe.

<u>Reading the Flame and the Soot of the Candle</u>

Setting up the spell and burning the candle is only half the work in candle spells; it's also important to interpret the spell. How the container or seven-day candle burns, which way the flame flickers, and what the soot left behind means are all good indications of how successful your spell work is going to be. Watching the flame burn is not only a divination and meditation tool but can reveal different meanings.

- If the candle doesn't light it means that the spell won't work. A different candle and a different intention are needed at this time.
- If there are two or more flames, it usually means that there are other entities - spirits, deities, or humans - involved, either positively or negatively.
- If the flame burns high, it means there is great power behind the spell and little resistance will be met. If it is calm, you are in the presence of a guiding spirit and the intention will come true.

- If the candle goes out on its own in the middle of the spell, the spirits cannot help you and the outcome has been decided in a negative way.

For candle spells, reading the soot that forms around the edges of container is also a great way to tell if your spell will be successful.

- If the soot stays at the top rim of the candle or stops in the middle as it burns, there was a negative energy that has been unblocked and the spell now works. Halfway down or less means the spell was successful.
- If the soot goes all the way down the length of the candle, things are preventing you from attaining your desires. Soot surrounding the bottom of the glass means that a protection spell might be required.
- White soot suggests spiritual connection - the spirits have heard your prayers and they are working.
- If there is soot on one side of the candle only it could be an incorrect spell.

Reading the Wax of the Candle

If you are using votive or chime candles in your spell work, the way the wax melts is another good indicator of whether your spell will be successful or not. This kind of divination is called *ceromancy*, involves reading which side of the candle the wax melts down the symbols or shapes it forms when it hardens.

- If the wax melts down the front of the candle this represents everything on the physical plane, including your physical health, your money, your relationships with people, and your home.

- If the wax melts down the back of the candle this represents everything non-physical in your life, including spirits, spirituality, your beliefs, your emotions, and your mental health.
- If the wax melts down the left side of the candle this is representative of the past. If the wax melts down the right side of the candle, this represents the future.

The way that the wax creates patterns indicates the success of your spell will go and the symbols you should look for in your life.

- **Acorns** – Positivity in health and future security
- **Anchor** – Good luck, prosperity, and security
- **Apple** – Vitality and health
- **Arrow** – News, communication, and travel
- **Baby** – Birth of a baby
- **Birds** – News from a distance and good omens
- **Crescent Moon** – New beginnings or opportunities
- **Crown** – Success and leadership
- **Eye** – A warning to look out for enemies and situations
- **Heart** – Love and relationships
- **Key** – Uncovering obstacles
- **Owl** – Knowledge and intelligence
- **Rabbit** – Journey, return from the past
- **Rings** – Marriage
- **Rose** – Self-love and happiness
- **Star** – Good luck and fortune
- **Horseman** – Stranger entering your life
- **Ladder** – Advancement of something

- **Lines** – A journey, which depends on whether it is long or short
- **Scissors** – Separation of some kind
- **Spider** – Spiritual calling, inner guidance
- **Sun** – Good health and positivity

I hope this chapter has helped you bring a little bit of magic into your home and helped you set up a sacred space. Regardless of where it is or what is in it, even if you can only decorate one particular area or room in your house, it is important that you set up a sanctuary that speaks to you. Hygge witchcraft is not just about the items in that room or that house – it's about the feelings they bring, so make sure that where you choose to practice has the positive vibrations you need in your work.

Courtney Hope

House Magic

Now that you've cleansed the energy in your home and created a sacred space for your practice, your mind should be starting to clear and a feeling of ease should begin to wash over you. You should be content and comfortable in your home and the intentions you have set for pursuing witchcraft, ready to focus on what is now important – manifesting the things you desire in your life.

There are many different spells and practices to do this depending on what you wish to manifest. The items you use can range significantly, and there really isn't any right or wrong way to practice being a Hygge Witch. However, there are some spells I would like to walk you through, as I feel they bring out the essence of being a Hygge Witch. You can use all of them, or none of them, but I also suggest you research other spells you wish to do outside of this book. With all spells, make sure the intent is clear and positive and practice with an open heart.

Draw a Sacred Bath

One of the most hygge thing you can do is draw yourself a sacred bath and languish in the relaxing scents. I often do this while I conduct my *tasseography*, or cup-reading, with the remains of a cup of tea or coffee or a glass of wine. Water is a healing and soothing element, and using a variety of herbs, flowers, and candles that align with your required intention during your bath is a great way to soak up their power. Some of my favourites are a vegan version of a milk and honey bath, its relaxing indulgence fit for Cleopatra herself.

Another bath that works especially well during the time of the full moon is a salt bath. Salt is very purifying and a natural way to relax and soothe stressed and tense muscles, but ensure you are using Epsom salts or sea salt in your bath – table salt is never recommended. Use essential oils and soaps to add to the decadence of your bath.

Spell Jars

Spell jars are an excellent place to start when it comes to witchcraft, as they are fairly simple to put together and can be created at whatever time of the moon feels right to you. Spell jars are a collection of objects that promote the manifestation you are trying to achieve sealed into a jar, with a corresponding candle lit on top of it to seal the intention.

For all spell jars you can follow the same method, just changing the ingredients. Gather your items, cleanse your area, and set your intentions for the spell work. Put your items purposefully in your jar, meditating on the reason and the intention you are creating. Seal your jar and light the corresponding candle on top (a votive works very well for these), sealing the intent.

To seal the spell jar, I like to light a short chime candle and create a seal of wax around the lid and the spaces between the lid and the jar, letting the wax fall where it may. If you choose to do this spell candle ritual, move the jar clockwise until the entirety of the lid and the sides are covered in wax. If you have any candle left over after sealing the spell jar, melt a little in the centre of the lid and stick the spell candle to the top of the jar until it burns out.

Display the jar on your altar – unless something specific must be done to the contents, highlighted below - until your spell jar has come to fruition.

Once you feel the spell jar has served its purpose, you may discard it. You must do so properly, by either burying it at a crossroads, leaving it in a cemetery, or throwing it in running water. If you do this, ensure the items in the jar and the wax you have used to seal it are non-toxic and will not cause harm to any plant or animal life. If you are unable to do this, you can discard the used spell jar by removing the contents and cleansing the items in Florida water, disposing of the jar and items by throwing them away safely or burying them to return to Mother Earth.

Although the method for creating spell jars is all the same, there are many different spell jars you can create for different intentions. These are some of my favourites:

Happy House Spell Jar

A perfect gift at a housewarming or to light in your own home to spread the warmth, this spell jar is the welcoming hug you need after a long day! Display it on your altar or in a central area, like where you place your keys when you return home for the day.

To make a Happy House Spell Jar you will need the following:

- A jar

- Personal items (a strand of hair, your spit, and other bodily fluids you may prefer to use)
- Written intent on a piece of parchment
- Herbs for a protective home, including cinnamon for abundance, nutmeg for awareness, salt for security, apple pieces for life.
- A golden candle

Money Spell Jar

If you are looking to manifest some wealth, the money jar spell is a wonderful way to ensure an abundance coming your way. Bury this spell jar under a basil tree at the front of your house to affirm success.

To make a money spell jar, you will need the following ingredients:

- A jar
- Personal items (a strand of hair, your spit, and other bodily fluids you may prefer to use)
- Written intent on a piece of parchment
- A money drawing tincture
- Coins
- Herbs that promote abundance such as cinnamon, ginger for hot, fast, explosive success, honeysuckle and thyme to attract, and even bay leaves with the specific amount of money you need written on them
- A green votive candle

Honey Spell Jar

The honey spell jar is a great way to sweeten up a situation, person, or event to make them be nicer to you. Do this ritual at the new moon and keep the jar for up to seven days on your altar, reigniting the intention once a day.

To make a honey spell jar you will need the following ingredients:

- A jar
- Sugar
- Raw honey or maple syrup
- A photo of the person or event, or a written intention and specifics as to why you would like to sweeten this situation or person.
- A yellow candle

Health Spell Jar

If you are like me, you might be struggling with finding the motivation to start working out or need a little reminder to eat healthier. This health spell jar is amazing! Keep the jar on your altar until you feel the benefits starting to kick in.

Items you need for a health spell jar include:

- A jar
- A personal item (a strand of hair, your spit, and other bodily fluids you may prefer to use)
- Written intent on a piece of parchment
- Herbs such as coriander, eucalyptus, ginger, rosemary, thyme and sage, all used to represent good health.

- A cinnamon stick for healing and good health.
- A crystal that represents good health like agate, amethyst, jade, or sunstone. Ensure these crystals are charged by the full moon.

Protection Spell Jar

This is a house protection spell that can be made and either hung or nestled by any door using a gold chain or a gold ribbon.

To make a spell jar for your protection you will need the following ingredients:

- A jar
- Sea Salt
- Cascarilla powder (this is made by crushing eggshells and is traditionally used to cleanse, purify, and protect from evil forces)
- Labradorite crystals, used to awake mystical abilities and increase healing.
- A gold candle, gold ribbon or chain to hang the spell jar up.

Love Spell Jar

You've seen nineties witchcraft TV shows that depict a character crushing hard on someone, casting a love spell that goes awry, haven't you? Heed that pop culture warning, as love spells are extremely strong! Ensure your intention is very clear for this love spell jar, and make sure you allow for free will to prevent any karmic ramifications.

To create a love spell jar, you will need the following ingredients:

- A jar
- White rose petals
- Basil essential oils
- Cinnamon sticks
- Sugar cubes
- Cleansed rose quartz crystals
- Your intention written on a piece of parchment

Spell Bags

Spell bags are similar to spell jars, but the magic is contained in a small bag you can carry with you. To make a spell bag, fill the bag with magically charmed items of your choosing like the ingredients from the spell jars above and close the bag tightly. The bag itself is part of the spell, so make sure to select the appropriate colour and material for the spell you are putting together.

Pocket Altars and Travel Altars

A pocket altar, like a spell bag, allows a witch to carry her intentions, spells, and magic with her wherever she goes. It's designed as a reminder of her overall goals in life and allows the elements she needs for her practice to come together at any time and any place. Carrying your pocket altar with you wherever you go means your magic goes with you everywhere! Pocket altars are easier to carry than a heavy jar - I like to suggest a vintage tin to keep your spell items, because they are cute and small enough to carry in your handbag!

Although they are similar, there is a difference between a pocket altar and a travel altar. A travel altar is composed of smaller items that you might need to conduct a spell in a pinch, or if you don't have all your spell items with you. Travel altars can include small candles (birthday candles work well here!) of any colour, oils or colognes, parchment and a pen, and even a small selection of basic crystals, like clear quartz, which works well as a substitute for any crystal specified in a spell. Whereas a pocket altar holds a spell.

Herbs and Flowers to Use in Spells

Utilising the herbs and flowers that correspond with your intentions in any of your spells is a good way to focus your manifestations. All witches should have a selection of herbs for their practices. Some of the handiest ones to have in your possession are:

- **Basil –** Protective qualities and a reputation for healing hurt romances. It also represents abundance and prosperity.
- **Bay Leaf –** Great for prosperity and abundance.
- **Chamomile –** Reduces anxiety and stress and can help you sleep.
- **Cinnamon –** Perfect for spells involving love, lust, healing, and purification.
- **Coriander –** Promotes healing, love, and lust.
- **Dandelion-** Promotes longevity, enhances psychic ability and is an emotional and spiritual cleanser.

- **Dill** – Good for protection, love, attraction, money, and strength.
- **Elderflower** – Promotes protection, beauty and divination.
- **Garlic** - Represents passion and strength.
- **Hibiscus** – Promotes love, lust, harmony and peace.
- **Juniper** – Great for cleansing, providing protection against accidents, and protection against illness.
- **Lavender** – Promotes healing, love, and happiness and heals grief and guilt. It's also an amazing sleep assistant.
- **Merigold** – Promotes the power to inspire.
- **Mint** – Uplifts your spirits and brightens your life's perspectives.
- **Nutmeg** – Promotes health, luck, and fidelity.
- **Pepper** – Provides protection and purification.
- **Rose** – Perfect for romance and love spells.
- **Rosemary** – Using rosemary wards off negative energy.
- **Saffron** – Boosts mood and helps promote wealth.
- **Sage** – Promotes healing, longevity, good health, and protection.
- **Thyme** – Provides purification and psychic cleansing and healing.
- **Witch Hazel** – Re-centres your energy.
- **Yarrow** – Represents marriage and courage, also great for hex breaking.

How you collect these herbs is completely up to you. Many Hygge Witches are adept at kitchen witchery and green witchery, so you can grow and harvest your own herbs in your back garden or use the spices and herbs from the kitchen rack.

It really doesn't matter where you get them from if the plant is sacred to you and harvested appropriately.

With these basic elements, you will be making magic in your kitchen in no time!

Divination

I love celebrating and spending time with the people I love over a glass of red wine, a roaring fire, and a selection of delicious vegan foods. Sharing a meal together is the very epitome of a hygge lifestyle. But you can also use time spent enjoying other people's company to conduct fortune telling and divination.

Divination is 'the practice of seeking knowledge of the future or the unknown by supernatural means.' There are many different forms that you can conduct a fortune reading. You can use tarot cards and affirmation cards, read the lines on your palms, or you can put your kitchen witchery skills to good use and read the sediments left over from many forms of drinks, including wine, coffee and the most popular option, tea leaves.

Wine Spell Correspondents

The benefits of using wine in witchcraft have not been widely discussed, but as both an avid red wine drinker and a believer in sitting around a fire drinking a glass of red wine as the ultimate hygge experience. I also strongly support the use of wine in ritualistic practices. Even Jesus incorporated red wine into the church's Sunday rituals!

Wine is associated with happiness, success, love, relationships, and offerings. Having a glass of wine present on your altar during a spell for these intentions will assist in manifesting your goal. The type of wine you offer is based on the different associations of different wines:

- **Cabernet Sauvignon –** This red wine is of the earth element due to its full-bodied alchemy. Use Cabernet Sauvignon in spells associated with grounding, protecting, banishing, strength building, energy, and fertility.
- **Merlot –** A slightly softer red, Merlot is a good wine to use in spells associated with unity, love, passion, self-care, protection, healing, sexuality spells and sea witchcraft. It is the elements of water and fire.
- **Pinot Noir –** This fresh, fruity red is the elements of earth and air, and can be used in spells associated with prosperity, protection, wealth, and success.
- **Shiraz –** This hearty red is a true-blue fire and earth element. It's perfect for spells focused perfect on wealth, banishing, divination, comfort, mystery and solving secrets.
- **Chardonnay –** This light, citrus-y white wine is associated with the element of water and is a good representative in spells involving peace, emotions, safety, success, happiness, balance, and purification.
- **Riesling –** The crisp, steely white is related to fire and water elements and works well in spells associated with energy, movement, growth, rebirth, love, friendship, and attraction and cleansing spells.
- **Sauvignon Blanc –** This herbal white evokes the elements of air and earth, associated with rituals that work on love, peace, friendship, companionships, healing, happiness, and spirits.
- **Rose –** This sweet, fruity white-red hybrid is used for the element of air which is perfect for spells that work

on new beginnings, happiness, excitement, friendship, new romances, and playfulness.

- **Sparkling** – Sweet sparkling wines like champagne work well in spells associated with success, completion, celebration, wealth, opportunities, prosperity, and space or weather witchcraft.

The art of reading the sediment at the bottom of a wine glass is called *Oenomancy*, and was traditionally performed by a priestess of Bacchus, the Roman god of wine. Oenomancy can be performed by spilling wine on cloth or paper and studying the resulting stains, or by reading the sediment at the bottom of the glass.

How to Read Tea Leaves

If you or your friends don't drink alcohol, you can use other drinks in your craft instead. Divination with tea is one of the most classic forms of future-telling, and sipping a delicious tea is an excellent way to cultivate hygge, so it makes sense to combine the two elements into your Hygge Witch practices. Reading tea leaves, or *Tasseography*, is easy. Although you can choose to use a special teacup to read tea leaves, like those with reading options etched into the ceramic, or you can read the leaves in any cup suitable for tea.

- To read tea leaves, pour a teaspoon of loose-leaf tea into a cup and steep them in boiling water. You can use flavoured tea leaves aligned with your intentions, like green tea for boosting energy and cleansing, chamomile for reducing anxiety and assisting in sleep, peppermint for clarity, or white tea for cleansing and protecting. When you use teas for divination, do not add any sugars or milks into your tea; they should be herbal only.

- Drink the tea, leaving a small amount of liquid at the bottom of the cup, before the person whose fortune is being read takes the cup by the handle in their left hand with the rim pointing upwards, rapidly moving the cup three times from left to right.

- Overturn the cup onto its saucer and allow the leaves and water to drain out. Put the cup back to right side up and read the placement of the leaves in the cup.

The cup is divided into three parts: the rim, the sides, and the bottom. Any leaves at the rim indicate what will occur in the present, those on the side are the near future and those on the bottom are the distant future. The nearer the icons are to the handle, the nearer the divination is likely to come true.

Coffee Fortune Telling

Less well known than reading tea leaves, you can also tell fortunes from coffee, a favourite past time on the streets of Turkey following the drinking of ground or percolated coffee. Like reading tea leaves, your fortune is told by the leftover coffee grounds in your cup; this is traditionally done by someone else, as it is not customary to read your own cup, making coffee-based fortune telling a slow, hygge moment of connectivity.

- To start a reading, prepare a strong coffee with ground coffee, ensuring to keep some of the grounds in with the coffee.
- Serve the coffee in a cup with a solid white on the inside, often known as a fortune telling cup.
- Ask the person getting the divination reading to drink with intent and reflect on their questions as they drink the coffee.
- When they are done drinking, turn the cup over upside-down on the saucer and swirl the cup and saucer 3 times clockwise. Allow the cup to sit for a few minutes to settle before conducting your reading.

In a past, present, and future reading the bottom of the cup represents the past, the midsection of the cup shows the present situation, and the area around the rim symbolises the future. The area around the handle of the cup represents love and relationships, the area across from the cup's handle is the section related to finances; the section right of the handle predicts future events, and the bottom of the cup advises family and home life.

<u>Tasseography Icon Meanings</u>

Just like candle wax, the leftover leaves from your tea or the sediment from wine or coffee forms shapes that can be interpreted in many different ways. If you see any of the following icons, there are meanings attached to them:

- **Anchor –** Good luck in business and a stable love life
- **Angel –** Good news, happiness and protection
- **Apples –** Long life and success in school or business

- **Arrow –** Directions
- **Bat –** Stealthy intentions
- **Bee –** New friends or friends gathering
- **Birds –** Good luck
- **Boats –** A visit from a friend will occur
- **Butterfly –** Success and pleasure are in your future
- **Car –** Approaching wealth
- **Cat –** Solitude or fights with friends
- **Coffin –** Unlike the Death tarot card, this does mean death!
- **Cross –** Trouble ahead by either delay or death
- **Dagger –** Sign of divorce
- **Dog –** Friendship and loyalty
- **Fish –** Goods news will be had from another country
- **Hat –** Success in life
- **Heart –** Good things are coming your way
- **Horseshoe –** Success in finding a partner
- **Hourglass –** Imminent danger!
- **Key –** Moving and new solutions
- **Kite –** Being tethered or tied down
- **Ladder** – Travel will be in your future
- **Lines** – A journey; the length of the line will indicate how long that journey will be for
- **Monkey** – Curiosity
- **Moon** – Happiness and success
- **Mushroom** – The sudden separation of lovers after a fight
- **Owl** – Warnings about unlucky events
- **Pear** – Wealth and financially beneficial situations
- **Rabbit** – You will have success in the city

- **Raven** – Bad or sad news
- **Ring** – This indicates marriage, but if an initial is nearby this will indicate the initial of the future spouses' name
- **Scissors** – An argument or a breakup
- **Snakes** – Snakes are bad omens! Take precaution in whatever you do
- **Spider** – Money is on the way to you
- **Star** – Good luck
- **Trees** – Good luck and prosperity
- **Triangles** – Changes are on the horizon
- **Umbrella** – Difficulty or annoyance in something
- **Wolf** – The need for family

Alongside using common household ingredients to read your fortune, you can create fabulous meals with the same power behind them through *Kitchen Witchery*.

Courtney Hope

Kitchen Witchery

Kitchen Witches conduct their magic by creating delicious meals. They infuse each recipe with intent, manifesting their desires into delicacies that can be enjoyed by everyone.

Remember that whatever you put into your kitchen witchery will be ingested or breathed in, so you need to make sure your ingredients are safe for human consumption to avoid any nasty surprises!

A good idea for the kitchen is including the herbs, fruits, nuts, and other edible ingredients that correspond to your intention in a specific spell manifested as a loaf of bread or a beautiful tart. If you aren't comfortable making up your own recipes to mix spell ingredients, I have provided some of my favourite Hygge Witch recipes to get you used to the idea of cooking with magical intention.

Vegan Spinach Creamed Dip

This recipe is a dish to break out at events and celebrations like Ostara, where sharing is caring! A cob loaf always goes down a treat and looks great with floral arrangements for spring.

Ingredients:

- 450 gms cob loaf
- 250 gms spinach
- 250 gms vegan creamed cheese
- 250 gms tofu
- 2 tbsp nutritional yeast
- 2 tbsp salt and pepper
- Crackers, carrot sticks, and celery to serve

Method:

1. Preheat oven to 180 degrees Celsius and line a large baking tray with baking paper.
2. Cut about 4 cms off the top of the cob loaf to form a lid. Hollow out the centre of the loaf by scooping out the bread, making sure to leave a thick rim around the sides and bottom of the loaf.
3. In a food processor, mix the tofu, vegan sour cream, nutritional yeast, salt, pepper, and spinach until smooth.
4. Spoon the dip mixture into the loaf of bread and replace the lid on the cob. Place the loaf and the spare bread taken from the interior onto the baking tray and put in the oven to bake for 20 minutes or until golden brown.
5. Remove from oven and allow to cool. Serve with celery, carrot sticks, bread, and crackers.

Herb-Infused oils and vinegars

Making herb-infused oils and vinegars is a great way to create an easy-to-use oil for your altar or to incorporate your spell herbs into recipes. Check out page 92 for information on different herbs and their spiritual properties.

Ingredients:

- Any type of herb you would like to use or focus your intentions on.
- Cooking oil, wine, cider, or rice wine vinegar
- A strainer and a secure jar.

Method:

1. Chop or bruise a handful of whatever fresh herbs you would like to infuse your tincture with and place them in a tight-fitting oil jar.
2. Fill the jar with your chosen liquid (you can use wine, cider or rice wine vinegar if you would prefer a different tincture, but I love a classic sunflower oil) and leave it in the fridge for two to three weeks, shaking occasionally.
3. Strain out the herbal matter left in it and taste the tincture, adding more herbs and repeating the process if it needs to be stronger.

Hot Cross Bunnies

Perfect to break out during your celebration of Ostara or Easter, these Hot Cross Bunnies taste delicious and look almost too cute to eat! Why have a normal hot cross bun when it could be bunny shaped?

Ingredients:

- 2 cups plain flour
- 1/8 cup caster sugar
- ¼ cup cocoa powder
- Pinch of salt
- ½ cup Dark Chocolate Chips
- ½ cup mixed fruits as desired
- 20 gms butter substitute
- 150 ml soy milk
- 1 vegan egg
- ¼ cup plain flour
- 2 tbsp water
- 1/3 cup water
- 2 tbsp sugar

Method:

1. Combine the flour, yeast, sugar, cocoa powder, salt, fruit, and chocolate chips together in a large bowl.
2. Put the butter in a small saucepan and melt on low heat. Pour in the milk into the saucepan and heat until it is lukewarm.
3. Add the milk and butter to the bowl with the dry ingredients and then add in the egg, beating the mixture until smooth. Don't over-knead the bread – it should be slightly sticky dough.
4. Cover the dough with cling wrap and leave in a warm dry spot in a bowl for about an hour to let the dough rise.

5. Once the dough has risen, split it into roughly 8 even sized balls. Roll them into the shape of an egg and with a pair of scissors, snip two bunny ears into the dough, making sure to cut about halfway back.

6. Using your fingers, shape the points to resemble bunny ears and use a toothpick or a skewer to poke holes in the dough just below the ears to create the eyes. Alternately, you can use icing sugar or chocolate chip cookies to make the eyes.

7. Place each bunny on the lined baking tray and leave them to sit for another half an hour.

8. Mix ¼ cup of flour and 2 tablespoons of water in a small bowl until smooth. Place the watery mixture into a small zip lock bag and once the dough has rested, cut the corner of the bag and pipe crosses over the backs of the bunnies.

9. Put the bunnies into the oven and bake for 20 – 25 minutes until cooked and a small skewer inserted into them comes out clean.

10. Meanwhile, in a small pot, mix the sugar and water for the glaze together over a slow heat until the sugar has dissolved.

11. Brush the glaze mixture over the top of the bunnies once out of the oven to give them a glossy look. Serve with spreads and jams once cool.

Mulled Apple Cider

This is the perfect drink to have during Mabon celebrations and all through the autumn hygge season. Apples are usually at the best during this season, and partaking in hygge activities such as apple picking is a perfect way to celebrate.

This fruity concoction has many magical uses. The oranges bring luck, while apples are the perfect ingredient for love and sex magic, fertility assistance, and immortality. Ginger helps you draw in new experiences, and cloves help you gain what you are seeking.

Ingredients:

- 1 gallon fresh apple cider or unfiltered apple juice
- 1medium orange
- 1 knob fresh ginger
- 1 tbsp whole cloves
- Orange slices for serving.

Method:

1. Cut the orange into thick rounds and cut the ginger into slices. Put them in the slow cooker.
2. Pour the cider or juice into a slow cooker. Add cinnamon sticks, cloves, and optional ingredients items such as the orange slices, or extra spiced rum for a more alcoholic kick.
3. Cover the slow cooker and keep it on low for 4 hours while the cider stews. Serve in Moscow Mule mugs!

<u>Mulled Wine</u>

This is a staple alcoholic beverage for Danish hygge winters and witchy events! Otherwise known as *Gluhwein*, there are many different variants of mulled wine. The perfect warm and comforting drink, often served at Christmas, Yule, and other celebrations in the wintery months. At the beginning of the season, I dry out thin slices of lemons, apples, and oranges in the oven at a low heat to keep in a jar for garnishing the whole season's worth of wine.

This recipe really highlights the intentions of the ingredients. Cloves represent protection and are often used in exorcisms, while cinnamon represents healing, protection, prosperity, and strength. Mixed with the perfect house blessings and luck of oranges, maple syrup for longevity and Star Anise for psychic awareness, drinking mulled wine is an excellent way to protect your home and those inside it.

Ingredients:

- 2 small oranges
- 1 bottle of red wine
- 1 - 2 tsp maple syrup
- 2 whole cinnamon sticks
- 3 star anises
- 4 whole cloves
- ¼ cup brandy
- Optional garnishes such as dried fruit slices, lemons, or cranberries.

Method:

1. Slice the oranges into rounds and place them in the bottom of a slow cooker, lightly juicing the oranges into the pot.
2. Pour wine, brandy and maple syrup into the pot and add the cinnamon sticks, star anise, and cloves.
3. Warm the mixture in the slow cooker until it reaches your preferred consistency. The longer you let it sit, the richer and more alcoholic your mulled wine will be. Add in any additional garnishes and flavours prior to serving.

Rosemary Baked Brie with Candied Walnuts

This recipe is another delicious treat for entertaining, and while my recipes are vegan, you can use a normal brie cheese wheel if you would prefer. This favourite recipe of mine came from the mind of Willow Arlen at *Will Cook for Friends*.

Rosemary is often used for good health, healing, and removing negativity, so this is a tasty way to manifest good health for yourself and your family. Walnuts are used to access divine energy and bring about blessings of the Gods, and maple syrup promotes longevity, money, and love.

Ingredients:

- 1.5 cups of walnuts
- ¼ cup maple syrup
- Pinch of salt
- 1 wheel of vegan brie style cheese
- 2 - 3 tbsp maple syrup or rice malt syrup
- A few fresh sprigs of rosemary

Method:

1. Pre-heat the oven to 170 degrees Celsius. Line a cookie sheet with baking paper and set it aside.

2. Heat up a pan on the stove without oil or butter. Add the walnuts, maple syrup and salt to the pot and stir constantly until the maple syrup starts to thicken and has attached to the walnuts.

3. Pour the caramelised nuts onto the prepared cookie sheet and spread them out, placing the cookie sheet into the fridge to set.

4. Remove the cheese from its packaging and set it in the centre of a separate oven tray. Place a sprig of rosemary on top and drizzle the maple or agave syrup on top of the cheese.

5. Place the cheese in the oven and bake for 10 – 12 minutes or until the cheese is soft to touch.

6. Remove from the oven and let the cheese cool slightly. Remove the candied walnuts from the fridge and serve on top of the cheese with crackers or rustic bread.

Sourdough Starter

Sourdough starter is a way to make your own sourdough bread without using any other ingredients other than flour and water. It's super easy, and nothing is more hygge then the smell of bread baking!

Ingredients:

- ½ cup water.

- 1 cup white flour

Method:

1. Using a glass measuring cup, mix the flour and water until the flour has dissolved. It should be a thick paste. Loosely place the lid of the jar or a wet towel onto of the measuring cup and let it sit at room temperature for 24 – 48 hours.
2. When the mixture starts to bubble, discard all but ½ cup of the starter. Combine the remaining starter, 1 cup of flour, and another ½ cup of water. Mix well, cover again and leave for another 24 hours. This is known as "feeding" the sourdough starter.
3. The next day there will be more bubbling; you will need to "feed" the sourdough starter again using step 2 twice a day until the starter peaks and starts sinking again. Once the starter has sunk, you need to feed it again - this can take a further 24 – 48 hours. Although you can often tell that sourdough starter is ready to be turned into bread when it starts bubbling, you can check to make sure it's ready by dropping a teaspoon of the mixture into a glass of water. If it floats, it's ready to go..

<u>Sourdough Bread</u>

Once you've made your sourdough starter, you are ready to get baking. I love this super simple recipe for making your own bread without a bread maker!

Ingredients:

- 3.5 cups white bread flour
- Pinch of salt
- 1 cup water
- Any seeds or herbs you would like to put in your bread
- 1/3 cup sourdough starter

Method:

1. Leave your sourdough starter out for 24 hours prior to baking with the starter.

2. Mix 3.5 cups of white bread flour and a pinch of salt. If you are adding in different herbs or cheese into the bread you can add these in here. Roasted garlic, chia seeds, dill, and nutritional yeast are delicious options, but you can also make your bread according to your herb spell correspondents.

3. Mix in 1/3 cup of sourdough starter with 1 cup of water and stir.

4. Mix the sourdough starter with the dry ingredients until it becomes a thicker dough. You will not need to overly knead the bread. Cover with a damp kitchen tea towel and leave for 15 minutes.

5. Do two sets of stretching and folding of the dough to help build the gluten. Do this twice, at least 15 minutes apart.

6. Cover the dough with the damp kitchen towel and let it rise on the kitchen counter between 8 – 12 hours. It is ready when it is slightly domed, springy, and a little jiggled.

7. Place a sheet of baking paper into a tin and put the domed dough into the baking pan. Sprinkle with a little

flour, or some more herbs. Place the dough in the fridge for one hour to make it easy to score, and heat up your oven to 180 degrees Celsius.

8. Place the tin in the oven and heat for about 22 minutes or until deeply golden and crispy. Allow the bread to cool and then serve.

Vegan Cheese Dip

A delicious vegan recipe perfect for sustainable and cruelty-free entertaining any time of the year. This very garlicky recipe, promotes healing and purification and guards against the envy and ill-intent of others. Mixed with the protective qualities of rosemary and the monetary symbolism of cashew nuts, this vegan cheese dip with have you manifesting luck and keeping it!

Ingredients:

- 1 cup soaked cashew nuts
- 1 clove garlic
- 2 tbsp Tapioca flour
- 2 tbsp salt
- 1.5 tsp nutritional yeast
- 2 tsp apple cider vinegar
- ¾ cup warm water
- Extra slices of garlic, finely chopped
- Rosemary
- Cooking oil

Method:

1. Pre-heat oven to 180 degrees Celsius.
2. Put all the ingredients into a food processor and blend until smooth.
3. Pour the smooth mixture into a saucepan on the stove and stir until the mixture is thickened but still a liquid.
4. Once the mixture has thickened, pour it into a greased or lined over-proof dish and put extra slices of garlic and rosemary on top. Add a few splashes of olive oil.
5. Put the mixture into the oven and bake for 20 minutes until the top is golden brown.
6. Remove from the oven and allow the dip to cool for 5 minutes before serving with rustic bread.

Being a witch is about learning to be kind and compassionate to Mother Nature and understanding the workings of the world around us. To understand Mother Nature deeply is to see her spirit in the creatures that also share this planet with us, so I have curated these recipes without animal products.

Courtney Hope

CONCLUSION

Being a Hygge Witch has become a very large part of my identity; most of the workings in this guidebook are conducted constantly throughout my own life. That is the beauty of Hygge Witchcraft – it is magic that can be done every day! Hygge witchcraft is about manifesting your desires, ensuring your hard work pays off, and keeping positive hygge-friendly vibes surrounding you throughout your practice.

My hygge home has become my ultimate sacred space. Only friends and loved ones of my choosing may walk through the doors, ensuring my world is welcoming and comforting to the right energies. By protecting and cleansing my house, I make sure that the vibrations in my home continue to operate at the frequency I need to work, conduct spells, and enjoy my daily routine. My sanctuary provides the sense of calm that comes from a tidy and well-maintained home: "*A clean house is a clean mind*"

Without being distracted by cluttered living, I am free to live slowly, choose to do what is important to me, and take the time to enjoy it. Now *that* is the true meaning of hygge!

Choosing what is important to you and how you choose to spend your time is a crucial part of being a witch. Witchcraft is about exploring new possibilities in an opportunity that may not have been there before, or that you were too distracted to see. The same goes for this book. I am so thankful that you spent your time reading it and I appreciate your support of Hygge Witches, but it is also crucial that you continue to work in practices that are right for you specifically.

It doesn't matter if you've read through this book and decided you find a better fit as a green witch or a kitchen witch, or even practicing a form of magic that hasn't been mentioned here. That's totally ok! You need to continue to move and practice where and how your gut instincts tell you, and while I hope this book has assisted you in creating that, only you can build your perfect Hygge Witch life.

So, whether it's a beige and white room with minimalistic perfection, an urban warehouse overflowing with lush green plants, or somewhere in between, remember that YOU are the one that creates the magic surrounding you and the things you manifest into your life. I hope with the extra knowledge you've gained, you can go forward and manifest the best life possible!

This is your life – only you can determine the energy surrounding you.

ABOUT THE AUTHOR

More titles from Courtney Hope:

Secrets of a Party Planner
Cosmic Decay: Contamination
Cosmic Decay: Debris
Cosmic Decay: Absolution

Courtney Hope is an author and interior designer hailing from Canberra, Australia. Living her life in pursuit of hygge – the Danish concept of contentment in your surroundings – Courtney specialises in house magic and re-aligning the energies of the home to create a welcoming and comforting space.

A dedicated Halloween Queen and horror movie lover, Courtney is the author of the horror/science fiction series *Cosmic Decay*, as well as having written an event planning handbook called *Secrets of a Party Planner*, and a dating and relationship guidebook published under a pseudonym. She was the creator of the event planning blog *The Party Connection* which helped to foster her passion for writing and design.

Courtney is an environmentally friendly vegan, a pop culture fan, and lives with her fur son, a Japanese Spitz named Buddy.

www.courtneyhope.com.au
www.hyggewitch.com.au

REFERENCES

Ankarloo, Bengt and Stuart Clark, eds. Witchcraft and Magic in Europe: The Period of the Witch Trials. Philadelphia: University of Pennsylvania Press, 2002.

Ankarloo, Witch Trials, 66.

Ann Murphy-Hiscock. Hearth witch

Cho. Anjie. How to Create Good Feng Shui In Your Home. The Spruce. 2019

Choc Chip Hot Cross Bunnies. Food and styling – Jono Fleming. Photography – Denise Braki. Temple and Webster. https://www.templeandwebster.com.au/style-and-advice/Hot-cross-buns-2-ways-E10711 27/03/2017

Clea Shearer and Joanna Teplin. The Home Edit Life. ISBN: 9781784727161. Published by Octopus. September 29 2020

Clean House, Clean Mind. Psychology Today. 2020

Dictionary definition

Dorte Johansen. Newsletter; 'The two key components for creating a relaxing home'.

Dorte Johanson. Interview with Courtney Hope. 15.08.2022

Dorte Johanson. Interview with Courtney Hope. 15.08.2022

Dorte Johanson. Interview with Courtney Hope. 15.08.2022

Feldmann, Erica. Hausmagick: Transform your Home with Witchcraft. HarperOne 2019

Global Wellness Institute. Opportunities and Impacts of Wellness for Regional Development

Global Wellness Institute. What is wellness?

Hector Garcia Puigcerver. Ikigai: The Japanese Secret to a Long and Happy Life. Hutchinson. 2018

House Magic: A Handbook to making every home a sacred sanctuary. Aurora Kane. Introduction.

Huffpost.com. June 23 2014. Nancy Graham Holm, Burning witches in Denmark: Midsummer ritual unintentionally endorses gender based violence.

Interior Design Institute Module Three: Design Styles. Version 1:1 July 2021

Interior Design Institute Module Three: Design Styles. Version 1:1 July 2021

Interior Design Institute Module Three: Design Styles. Version 1:1 July 2021

Kate Lohnes. How Rye Bread May Have Cause The Salem Witch Trials. Britannica.com.

Larsen and Eriksen Cophenhagen. The History and Concept of Minimalism.

Malleus Maleficarum. Encyclopaedia Britannica Kraemer and Sprenger. 1998.

Marie Kondo. The KonMari Method. https://konmari.com/about-the-konmari-method/

Marie Kondo. The KonMari Method. https://konmari.com/about-the-konmari-method/

Merriam Webster Dictionary.

Montague Summers (4 April 2014) [1926]. The History of Witchcraft and Demonology (Reprint ed.). Routledge. p. 66. ISBN 978-1-317-83267-6.

Pamer. Kerrilynn. Build an Alter to Bring Your Inspiration and Intentions to Life. The Chalkboard. 2018.

Sacred Smoke: The use of herb incense in North European and other Indigenous traditions. Scandinavian Center for Shamanic Studies. First printed in Sacred Hoop Magazine in issue 24, 1999. Nicholas Breeze Wood of Sacred Hoop.

Weaver, Dr. Libby. Women's Wellness Wisdom. Little Green Frog Publishing. 2016

Wiking. Meik. The Little Book of Hygge. Penguin Books Ltd. 2016.

Wiking. Meik. The Little Book of Hygge. Penguin Books Ltd. 2016.

Willow Arlen. Baked Brie with Rosemary, Honey and Candied Walnuts. 07/01/2014. Will Cook for Friends - https://www.willcookforfriends.com/2014/01/baked -brie-with-rosemary-honey-candied-walnuts.html

Witches: Real Origins, Hunts and Trails.
www.history.com. 12 Sept 2017

Milton Keynes UK
Ingram Content Group UK Ltd.
UKHW021547160924
1673UKWH00057B/337

9 781088 186664